Solutions Manual

for Atkins's

The Elements of Physical Chemistry

SECOND EDITION

Dixie Goss
Hunter College
City University of New York

W. H. Freeman and Company
New York

ISBN: 0-7167-3068-5

Printed in the United States of America

First printing 1997, VB

Contents

Preface

This manual contains solutions and answers to the problems in Atkins's, *The Elements of Physical Chemistry,* Second Edition. Additional problems are located at the end of each chapter. The solutions to these problems are found at the end of this manual. I have tried to provide a guide to working the problems as well as numerical answers. It will not help to simply read the *Solutions Manual.* The best way to understand physical chemistry is to do as many problems as possible. Working through these problems will help you to develop an understanding of the subject matter in a way that reading the solutions cannot. I would like to thank Peter Atkins for his assistance and valuable suggestions. Any remaining errors in the problems are mine.

D. J. G.

Chapter 1

States of matter and the properties of gases

1.1 (a) $110 \text{ kPa} \times \frac{760 \text{ torr}}{101.325 \text{ kPa}} = \underline{\mathbf{825 \text{ torr}}}$

(b) $0.997 \text{ bar} \times \frac{100 \text{ kPa}}{1 \text{ bar}} \times \frac{1 \text{ atm}}{101.325 \text{ kPa}} = \underline{\mathbf{0.984 \text{ atm}}}$

(c) $2.15 \times 10^4 \text{ Pa} \times \frac{1 \text{ kPa}}{10^3 \text{ Pa}} \times \frac{1 \text{ atm}}{101.325 \text{ kPa}} = \underline{\mathbf{0.212 \text{ atm}}}$

(d) $723 \text{ torr} \times \frac{101.325 \text{ kPa}}{760 \text{ torr}} \times \frac{10^3 \text{ Pa}}{1 \text{ kPa}} = \underline{\mathbf{9.64 \times 10^4 \text{ Pa}}}$

1.2 $2.045 \text{ g} \times \frac{1 \text{ mol } N_2}{28.02 \text{ g}} = 0.07298 \text{ mol}$

$$P = \frac{nRT}{V} = \frac{(0.07298 \text{ mol})(8.3745 \text{ kPa K}^{-1} \text{ mol}^{-1})(294 \text{ K})}{2.00 \text{ L}}$$

$= \underline{\mathbf{89.2 \text{ kPa}}}$

1.3 Arranging the perfect gas law in a form to solve for pressure:

$$p = \frac{nRT}{V}$$

$n = \frac{0.255 \text{ g}}{20.18 \text{ g mol}^{-1}} = 1.26 \times 10^{-2} \text{ mol}, \; T = 122 \text{ K},$

$V = 3.00 \text{ L}$

Therefore,

$$p = \frac{(1.26 \times 10^{-2}\ \text{mol}) \times (0.082058\ \text{L atm K}^{-1}\text{mol}^{-1}) \times 122\ \text{K}}{3.00\ \text{L}}$$

= **4.22×10^{-2} atm**

1.4 $$n = \frac{PV}{RT} = \frac{(24.1\ \text{kPa})\left(250.0\ \text{mL} \times \frac{1\ \text{L}}{10^3\ \text{mL}}\right)}{(8.3745\ \text{kPa K}^{-1}\ \text{mol}^{-1})(292.6\ \text{K})}$$

= 0.00248 mol = **2.48×10^{-3} mol**

1.5 1.04 kg - 0.74 kg = 0.30 kg mass of CO_2

0.300 kg x 10^3 g/kg = 300 g CO_2

$$300\ \text{g} \times \frac{1\ \text{mol}}{44.01\ \text{g}} = 6.82\ \text{mol}$$

$$P = \frac{nRT}{V} = \frac{(6.82\ \text{mol})(8.3745\ \text{kPa K}^{-1}\ \text{mol}^{-1})(293\ \text{K})}{2.50\ \text{L}}$$

= **6.64×10^3 kPa**

1.6 Remember that p is inversely proportional to V. Therefore

$p_2 = \frac{V_1}{V_2} \times p_1$. Converting 1.0 L to cm^3 gives 1.0×10^3 cm^3.

p_1=1.00 atm

$$p_2 = \frac{1.0 \times 10^3\ \text{cm}^3}{100\ \text{cm}^3} \times 1.00\ \text{atm} = 10 \times 1.00\ \text{atm}$$

= **10 atm**.

1.7 $\frac{P_1}{T_1} = \frac{nR}{V} = \frac{P_2}{T_2}$

$$P_2 = \frac{(973\ \text{K})(125\ \text{kPa})}{(291\ \text{K})} = \underline{\mathbf{418\ kPa}}$$

1.8 $P_1V_1 = P_2V_2$

$$P_2 = \frac{(7.20\ \text{L})(125\ \text{kPa})}{(4.21\ \text{L})} = \underline{\mathbf{173\ kPa}}$$

1.9 $\frac{T_1}{V_1} = \frac{T_2}{V_2}$

$$T_2 = \frac{(0.100\ \text{L})(295.3\ \text{K})}{(1.00\ \text{L})} = \underline{\mathbf{29.5\ K}}$$

1.10 Pressure is constant, so the same relation between temperature and volume as in the previous problem is used. Here the volume has increased by 14%, so $V_2 = 1.14V_1$.
$T_1 = 340$ K.

$$T_2 = \frac{1.14\,V_1}{V_1} \times 340\ \text{K} = 1.14 \times 340\ \text{K} = \underline{\mathbf{388\ K}}.$$

1.11 (a) $\frac{P_1V_1}{T_1} = \frac{P_2V_2}{T_2}$

$$V_2 = \frac{(104\ \text{kPa})(2.0\ \text{m}^3)(268.2\ \text{K})}{(294.3\ \text{K})(52\ \text{kPa})} = \underline{\mathbf{3.6\ m^3}}$$

(b) $V_2 = \dfrac{(104\ \text{kPa})(2.0\ \text{m}^3)(221.2\ \text{K})}{(294.3\ \text{K})(0.880\ \text{kPa})} = \mathbf{\underline{178\ m^3}}$

1.12 $V = \dfrac{n_J RT}{p_J}$

(a) There is only one volume, so using the amount of neon and p_J for neon we can calculate the volume.

$n(\text{Ne}) = \dfrac{0.225\,\text{g}}{20.18\,\text{g mol}^{-1}} = 1.12 \times 10^{-2}$ mol,

p(Ne) = 15.2 kPa, T = 300K

$V = \dfrac{(1.12 \times 10^{-2}\ \text{mol}) \times (62.36\,\text{L Torr K}^{-1}\text{mol}^{-1}) \times 300\text{K}}{114\ \text{Torr}} = \mathbf{\underline{1.83\ L}}$

$p = \dfrac{nRT}{V}$ where n is the sum of the amounts of each component; therefore

$n = n(CH_4) + n(\text{Ar}) + n(\text{Ne})$

$n(CH_4) = \dfrac{0.320\,\text{g}}{16.04\,\text{g mol}^{-1}} = 2.00 \times 10^{-2}$ mol

$n(\text{Ar}) = \dfrac{0.175\,\text{g}}{39.95\,\text{g mol}^{-1}} = 4.38 \times 10^{-3}$ mol

$n = (2.00 + 0.438 + 1.12) \times 10^{-2}$ mol

$= 3.548 \times 10^{-2}$ mol

Substituting into the equation for p

$p = \dfrac{(3.548 \times 10^{-2}\ \text{mol}) \times (62.36\,\text{L Torr K}^{-1}\text{mol}^{-1}) \times 300\text{K}}{1.83\text{L}}$

= 363 Torr = **<u>48.4 kPa</u>**

1.13 To calculate the molar mass of the compound, we need to relate density, temperature and pressure. Using the perfect gas law:

$$pV = nRT \text{ and density} = \frac{\text{mass}}{\text{volume}}$$

$V = \frac{nRT}{p}$ and $\frac{m}{\rho} = \frac{nRT}{p}$, $\frac{m}{n} = \frac{\rho RT}{p}$. Note that R must be in the same units as T and p;in this case K and Pa.

$$M = \frac{(1.23 gL^{-1})(8.3145x10^{3} PaLK^{-1}mol^{-1})(330)}{(2.55x10^{4} Pa)}$$

$= \underline{\mathbf{132\ g\ mol^{-1}}}$

1.14 $V_m = \frac{V}{n}$ using the perfect gas law to calculate n,

$n = \frac{pV}{RT}$. Then converting units in order to use

$R = 8.2056 \times 10^{-2}$ L atm $mol^{-1}K^{-1}$

$V = 250\ cm^3 = 250$ mL $= 2.50 \times 10^{-1}$ L

p = 152 Torr = 152 Torr x 1 atm/760 Torr
= 0.200 atm

T = 298 K, substituting into the perfect gas law written above

$$n = \frac{(0.200\ atm) x (0.250 L)}{(8.2056x10^{-2} L\ atm mol^{-1} K^{-1}) x 298\ K}$$

$= 2.04 \times 10^{-3}$ mol

The mass to the gas is given, 33.5 mg, so the molar mass,

$M = \frac{0.0335\ g}{2.04x10^{-3} mol} = \underline{\mathbf{16.4\ g\ mol^{-1}}}$.

1.15 a) Perfect gas $p = \frac{nRT}{V}$ for i) n = 1.0 mol,
$R = 8.2056 \times 10^{-2}$ L atm $mol^{-1}K^{-1}$

T = 273.15 in 22.414 L . Standard pressure is **<u>1 atm</u>**.
Substituting in

$$p = \frac{(1.0\ mol) x (8.0256 x 10^{-2} L\ atm K\ mol^{-1}) x 273.17 K}{22.414\ L}$$

Using a similar substitution for 1000 K and 100 cm^3

$$p = \frac{(1.0\ mol) x (8.0252 x 10^{-2} L\ atm K\ mol^{-1}) x 1000 K}{0.100\ L}$$

= **<u>8.2×10^2 atm</u>**

b) For a van der Waals gas, using Table 1.5, $a = 5.489\ L^2\ atm\ mol^{-2}$, $b = 0.0638\ L\ mol^{-1}$ the van der Waals equation is

$p = \frac{nRT}{V - nb} - a\frac{n^2}{V}$ For conditions i) substitution gives

$$p = \frac{(1.0 mol) x (8.0256 x 10^{-2} L\ atm\ K^{-1}\ mol^{-1})}{22.414\ L - (1.0\ mol) x (0.0638 L\ mol^{-1})}$$

$$-5.489\ L^2\ atm\ mol^{-2} \frac{1.0\ mol^2}{22.414\ L}$$

= **<u>0.99 atm</u>**

At conditions ii) p = **<u>1.7×10^3 atm</u>**

1.16 a) The partial pressure of the gases can be related to their mole fractions. The total amount is found by
$n = n(H_2) + n(N_2) = 2.0$ mol + 1.0 mol = 3.0 mol

$$x(H_2) = \frac{2.0\,\text{mol}}{3.0\,\text{mol}} = 0.67$$

$$x(N_2) = \frac{1.0\,\text{mol}}{3.0\,\text{mol}} = 0.33$$

$$p_J = n_J \frac{RT}{V}$$

Solving for $\frac{RT}{V}$

$$\frac{RT}{V} = \frac{(8.2058 \times 10^{-2}\,\text{L atm K}^{-1}\text{mol}^{-1}) \times (273.15\,\text{K})}{22.4\,\text{L}}$$
$$= 1.00 \text{ atm mol}^{-1}$$

$p(H_2) = 2.0 \text{ mol} \times 1.00 \text{ atm mol}^{-1} =$ **2.0 atm**

$p(N_2) = 1.0 \text{ mol} \times 1.00 \text{ atm mol}^{-1} =$ **1.0 atm**

b) the total pressure is the sum of the partial pressures
$p = p(H_2) + p(N_2) = 2.0 \text{ atm} + 1.0 \text{ atm} =$ **3.0 atm**

1.17 Pressure is inversely proportional to volume. There will be an increase in pressure due to the depth submerged, which can be calculated from $p = \rho g h$

$$V_f = \frac{p_i}{p_f} V_i$$

and Total pressure $p_i = 1.0$ atm
$p_f = 1.0 \text{ atm} + \rho g h$

$\rho g h = (1.025 \times 10^3 \text{ kg m}^{-3}) \times (9.81 \text{ m s}^{-2}) \times 50 \text{ m}$
$= 5.0 \times 10^5$ Pa

therefore

$p_f = (1.01 \times 10^5\ \text{Pa}) + (5.0 \times 10^5\ \text{Pa})$

$= 6.0 \times 10^5\ \text{Pa}$

$$V_f = \frac{1.01 \times 10^5\ \text{Pa}}{6.0 \times 10^5\ \text{Pa}} \times 3\,\text{m}^3 = \underline{\mathbf{0.5\ m^3}}$$

1.18 External pressure is p_i and pressure at the foot of the column filled with liquid is $p_i + \rho gh$. At equilibrium the two pressures are equal, so

$$p_f = p_i + \rho gh$$

$$p_f - p_i = \rho gh$$

$= (1.0 \times 10^3\ \text{kg m}^{-3}) \times (9.81\ \text{m s}^{-2}) \times (0.15\ \text{m})$

$= \underline{\mathbf{1.5 \times 10^3\ Pa\ (= 1.5 \times 10^{-2}\ atm)}}$

1.19 $pV = nRT$, with n constant

$$\frac{p_f V_f}{T_f} = \frac{p_1 V_i}{T_i} \quad \text{and solving for } p_f$$

$$p_f = \frac{p_1 V_i T_f}{V_f T_i} \quad \text{since } V = \frac{4}{3}\pi r^3$$

$$p_f = \left(\frac{r_i}{r_f}\right)^3 \frac{T_f}{T_i} \times p_i$$

$$p_f = \left(\frac{1.0\,\text{m}}{3.0\,\text{m}}\right)^3 \times \left(\frac{253\,\text{K}}{293\,\text{K}}\right) \times 1.0\,\text{atm}$$

$= \underline{\mathbf{3.2 \times 10^{-2}\ atm}}$

1.20 For an Ideal gas

$$P = \frac{nRT}{V}$$

$$10.00\ g\ CO_2\ x\ \frac{1\ mol\ CO_2}{44.01\ g}\ 0.2272\ mol$$

$$P = \frac{(0.2272\ mol)(8.3745\ kPa\ K^{-1}\ mol^{-1})(298.1\ K)}{0.100\ L} = \underline{\mathbf{5631\ \ kPa}}$$

For a van der Waals gas:

$$P = \frac{nRT}{V - nb} - a\left(\frac{n}{V}\right)^2$$

a = 3.59 atm mol^{-2} L^2

b = 0.043 L mol^{-1}

$$P = \frac{(0.2272\ mol)(0.08206\ L\ atm\ K^{-1}\ mol^{-1})(298.1\ K)}{(0.100\ L - (0.2272\ mol)(0.043\ L\ mol^{-1}))} -$$

$$3.59\ atm\ mol^{-2}\ L^2\left(\frac{0.2272\ mol}{0.100\ L}\right)^2$$

= 43.1 atm x 101.325 kPa/atm = **<u>4363 kPa</u>**

1.21 Writing the expression in terms of p

$$p = \frac{RT}{V_m - b} - \frac{a}{V_m^2}$$

$$= \frac{RT}{V_m 1 - \frac{b}{V_m}} - \frac{a}{V_m^2}$$

$$= \frac{RT}{V_m} 1 + \frac{b}{V_m} + \frac{b^2}{V_m^2} + \mathrm{L} \; - \frac{a}{V_m^2}$$

$$\text{because } \frac{1}{1-x} = 1 + x + x^2 + \cdots$$

$$= \frac{RT}{V_m} 1 + b - \frac{a}{RT} \frac{1}{V_m} + \frac{b^2}{V_m^2} + \mathrm{L}$$

Compare this expression with

$$p = \frac{RT}{V_m} 1 + \frac{B}{V_m} + \frac{C}{V_m^2} + \mathrm{L}$$

Therefore

$$B = b - \frac{a}{RT} \quad \text{and} \quad C = b^2$$

Since C = 1200 $cm^6\ mol^{-2}$,
b = $C^{1/2}$ = **<u>34.6 cm^3 mol^{-1}</u>**

a = RT(b-B)

= (8.206 x 10^{-2}) x (273 L atm mol^{-1}) x (34.6 + 27.7) cm^3 mol^{-1}

= (22.40 L atm mol^{-1}) x (56.3 x 10^{-3} L mol^{-1})

= **<u>1.26 L^2 atm mol^{-2}</u>**

1.22 From the previous exercise

$$B = b - \frac{a}{RT}$$

B = 0 when $\frac{a}{RT} = b$ or $T = \frac{a}{bR}$

For CO_2

$$T = \frac{3.59 \text{ atm mol}^{-2}\text{ L}^2}{0.043 \text{ L mol}^{-1} \text{ x } 0.08206 \text{ L atm K}^{-1}\text{ mol}^{-1}}$$

$$= \underline{\mathbf{1.02 \times 10^3\ K}}$$

1.23 The root mean square speed (r.m.s) is given by the formula

$$c = \sqrt{\frac{3RT}{M}}$$

$$c = \sqrt{\frac{3 \text{x} 8.314 \text{JK}^{-1}\text{mol}^{-1} \text{ x T}}{4.00 \text{g mol}^{-1}}}$$

$$c = 78.97 \text{ m s}^{-1} \text{ x} \left(\frac{T}{K}\right)^{1/2}$$

For T = 77 K, c = **693 m s^{-1}**

T = 298 K, c = **1363 m s^{-1}**

T = 1000 K, c = **2497 m s^{-1}**

For part b) substitute the molar mass of CH_4 (16.04 g mol^{-1})

$$c = 39.43 \text{ m s}^{-1} \text{ x} \left(\frac{T}{K}\right)^{1/2}$$

At T = 77 K, c = **346 m s^{-1}**

T = 298 K, c = **682 m s^{-1}**

T = 1000 K, c = **1249 m s^{-1}**

1.24 The formula for determining the mean free path is

$$\lambda = \frac{RT}{\sqrt{2} N_A \sigma p} \quad \text{solving for p,}$$

$$p = \frac{RT}{\sqrt{2} N_A \sigma \lambda}$$

with λ = 10 cm

$$p = \frac{(8.3145\,JK^{-1}mol^{-1})x(298.15K)}{\sqrt{2}\,x(6.03x10^{23}\,mol^{-1})x(0.36x10^{-18}\,m^2)x(0.10m)}$$

$= \underline{\mathbf{0.081\ Pa}}$

1.25 Solving for p with $\lambda \approx \sigma^{1/2} = \sqrt{0.36nm^2}$

$$p = \frac{RT}{\sqrt{2} N_A \sigma \lambda}$$

$$p = \frac{(8.3145\,JK^{-1}mol^{-1})x(298.15K)}{\sqrt{2}\,x(6.03x10^{23}\,mol^{-1})x(0.36x10^{-18}\,m^2)^2}$$

$= \underline{\mathbf{1.3\ x\ 10^7\ Pa}}$

1.26 The formula for determining the mean free path is

$$\lambda = \frac{RT}{\sqrt{2} N_A \sigma p}$$

$$= \frac{(8.3145\,J\,K^{-1}mol^{-1})x(217K)}{\sqrt{2}\,x(6.03x10^{23}\,mol^{-1})x(0.43x10^{-18}\,m^2)}$$

$$\times \frac{1}{(0.05\,\text{atm})\times(1.013\times10^{5}\,\text{Pa atm}^{-1})}$$

$= 970$ nm = **<u>1mm</u>**

1.27 The collision frequency is determined by the relation

$$z = \frac{\sqrt{2}N_A \sigma c p}{RT} \quad \text{with}$$

$$c = \sqrt{\frac{3RT}{M}}$$

$$z = 2^{1/2} N_A \sigma \sqrt{\frac{3RT}{M}} \frac{p}{RT}$$

$$z = 2^{1/2} \times (6.02 \times 10^{23}\,\text{mol}^{-1}) \times (0.36 \times 10^{-18}\,\text{m}^2) \times$$

$$\sqrt{\frac{3 \times (8.3145\,\text{JK}^{-1}\text{mol}^{-1}) \times 298\text{K}}{(39.95\,\text{gmol}^{-1})}} \times$$

$$\frac{10 \times 10^{5}\,\text{Pa}}{(8.3145\,\text{JK}^{-1}\text{mol}^{-1}) \times (298\text{K})}$$

= <u>5×10^{10} s^{-1}</u> so for 1 s, the number of collisions

is **<u>5×10^{10}</u>**

b) Substituting in other pressures gives **<u>5×10^{9} collisions</u>**

c) **<u>5×10^{4}</u>**

1.28 We have already calculated the number of collisions for a single Ar atom. Therefore we need to calculate the total number of atoms to determine the total number of collisions.

Since $n = \frac{PV}{RT}$,

the total number of molecules will be
$6.02 \times 10^{23}\ mol^{-1} \times n$

Number of molecules

$$= \frac{(6.02 \times 10^{23}\ mol^{-1})PV}{RT}$$

$$= \frac{(6.02 \times 10^{23}\ mol^{-1}) \times (10.1 atm) \times (1.0 L)}{(0.08206\ L atm K^{-1} mol^{-1}) \times (298 K)}$$

$= 2.486 \times 10^{23}$

Total number of collisions = $\frac{2.49 \times 10^{23} \times 5 \times 10^{10}}{2}$

= **$\underline{6.1 \times 10^{33}}$**

The answer is divided by two since two molecules collide in one collision event.

By substituting in for the other pressures,

b) **$\underline{6 \times 10^{31}}$ collisions**

c) **$\underline{6 \times 10^{21}}$ collisions**

1.29 $z = \frac{c}{\lambda}$ We have already calculated l in problem 1.20
To calculate c use

$$c = \sqrt{\frac{3RT}{M}}$$

$$= \left(\frac{3 \times (8.3145\, \mathrm{JK^{-1} mol^{-1}}) \times (217\, \mathrm{K})}{(28.02 \times 10^{-3}\ \mathrm{kg\ mol^{-1}})} \right)^{1/2}$$

[because 1 joule = 1 kg m^2 s^{-2}]

= 439 m s^{-1}

$$z = \frac{439\, \mathrm{m\, s^{-1}}}{(1000\, \mathrm{nm}) \times (10^{-9}\, \mathrm{nm/m})}$$

= **4×10^8 collisions/s**

1.30 $\lambda = \dfrac{RT}{\sqrt{2} N_A \sigma p}$; σ is given as 0.43 nm^2

Since we want to solve for several different p, calculate first in terms of p so it is easier to substitute into the equation

$$\lambda = \frac{(8.3145\, \mathrm{JK^{-1} mol^{-1}}) \times (298.15\, \mathrm{K})}{\sqrt{2} \times (6.02 \times 10^{23}\, \mathrm{mol^{-1}}) \times (0.43 \times 10^{-18}\, \mathrm{m}) \times p}$$

$$\lambda = \frac{6.8 \times 10^{-3}\, \mathrm{m}}{p / \mathrm{Pa}}$$

(a) When p = 10 bar = 1.0×10^6 Pa

$$\lambda = \frac{6.8 \times 10^{-3}\, \mathrm{m}}{1.0 \times 10^{6}} = 6.8 \times 10^{-9}\ \mathrm{m} = \mathbf{6.8\ nm}$$

(b) When p = 103 kPa

$$\lambda = \frac{6.8 \times 10^{-3}\,m}{103 \times 10^{3}} = 0.066 \times 10^{-6}\ m$$
$$= 6.6 \times 10^{-8}\ m = \underline{\mathbf{68\ nm}}$$

(c) $p = 1$ Pa $l = 6.8 \times 10^{-3}$ m = **<u>7 mm</u>**

1.31 The Maxwell distribution of speeds is

$$f = 4\pi\left(\frac{M}{2\pi RT}\right)^{3/2} s^2 e^{-Ms^2/2RT}\Delta s$$

At the center of the range, $s = 295\ m\ s^{-1}$

$$f = 4 \times \pi \times \left(\frac{28.02 \times 10^{-3}\ kg mol^{-1}}{2 \times \pi \times (8.3145 kg m^2 s^{-2} K^{-1} mol^{-1}) \times 500 K}\right)^{3/2} \times (295 m s^{-2})^2 e^{-(28.02)(295)^2/2 \times 8.31 \times 10^3 \times 500} \times 10 m s^{-1}$$

$= 9.063 \times 10^{-3}$ corresponding to **<u>0.91 %</u>**

1.32 $\lambda = \dfrac{RT}{\sqrt{2} N_A \sigma p}$

At constant V, p varies directly with T

$$p = \frac{nRT}{V}$$
and

$$\lambda = \frac{RT}{\sqrt{2} N_A \sigma} \times \frac{V}{nRT} = \frac{V}{\sqrt{2} N_A \sigma n}$$

λ **is <u>independent</u> of temperature.**

Additional problems

1.1 A sample of gas occupies 0.5 L at 100 K and 1 atm. What temperature is needed to expand the gas to 1.0 L (pressure constant)?

1.2 What pressure is exerted by 28.02 g of nitrogen in a 1 L container at 298 K if a) it behaves as a perfect gas b) it behaves as a van der Waals gas.

1.3 An unknown gas exerts a pressure of 2 atm at 298 K in a 1 L container. How many moles of gas are present?

1.4 A perfect gas occupies 1.0 L at 100 K and 1 atm. What is the pressure at 500 K and 2.0 L ?

1.5 The molar mass of a gaseous compound is 132 g mol^{-1}. What is its density at 330 K and 25.5 kPa?

1.6 What is the mean free path of argon at 200 K and 100 mPa? (s = 0.36 nm^2)

1.7 How many collisions does a single Ar atom make in 1.0 s when the pressure is 1.0 Pa and the temperature is (a) 298 K (b) 600 K and (c) 10 K? (s = 0.36 nm^2).

1.8 Calculate the volume of 1.0 mol NH_3 at 500 K and a pressure of 10 Pa assuming (a) the gas is perfect and (b) the gas is a van der Waals gas (a = 4.17 L^2 atm mol^{-2}, b = 0.037 L mol^{-1})

1.9 A sample of 255 mg of neon occupies 1.00 L at 200 K. Use the perfect gas law to calculate the pressure of the gas.

1.10 The density of a gaseous compound was found to be 2.46 g L^{-1} at 300 K and 25.5 Pa. What is the molar mass of the compound?

1.11 Calculate the mean free path of N_2 molecules in air ($s = 0.43\ nm^2$) at 25° C and (a) 100 kPa and (b) 1000 kPa.

1.12 A 100.0 g sample of dry gas is 32.0 G N_2, 48.0 g O_2 and 20.0 g Ar. What is the partial pressure of each component at (a) 100 kPa total pressure, (b) 400 kPa total pressure?

1.13 Calculate tha partial pressures of a sample containing 2.0 g N_2, 6.4 g CO_2, and 2.5 g O_2 at 90 kPa.

1.14 Use the Maxwell distribution of speeds to estimate the fraction of N_2 molecules at 500 K that have speeds in the range 270-290 m s^{-1}.

Chapter 2

Thermodynamics: the First Law

2.1 To calculate the work use the formula w = mgh

(a) w = 1.0 kg x 9.81 m s^{-2} x 10 m = **98 J**

(b) w = 1.0 kg x 1.60 m s^{-2} x 10 m = **16 J**

2.2 w = mgh

= 65 kg x 9.81 m s^{-2} x 4.0 m = **2.6 kJ**

2.3 w = mgh

mass is volume of column x density

v = p x r^2 x h

= 3.14 x $(0.50\ cm)^2$ x 760 mm x 1 cm/10 mm = 59.7 cm^3

m = 59.7 cm^3 x 13.6 g cm^{-3} = 812 g

To calculate work assume raising center of mass to 1/2 760 mm = 380 mm.

w = 380 mm x 1m/10^3 mm x 812 g x 1 kg/10^3 g x 9.81 ms^{-2}

= 3.03 kgm^2s^{-2} = **3.03 J**

2.4 w = $-p_{ex}DV$

To calculate p_{ex} convert to Pa

p_{ex} = 1.00 x 10^5 Pa

DV = 100 cm^2 x 10 cm = 1.0 x 10^3 cm^3 = 1.0 x 10^{-3} m^3

w = -1.00 x 10^5 Pa x 1.0 x 10^{-3} m^3 = -1.0 x 10^2 Pa m^3

Then , because 1 Pa m^3 = 1 J

w = **-1.0 x 10^2 J**

2.5 a) horizontally

work done by system = distance x force

= h x p_{ex} A = 155 cm x 105 kPa x 55.0 cm^2 x 1 L/10^3 cm^3

= **895 J**

b) vertically

w = hx m x g
= 155 cm x 1 m/100 cm x 250 g x 1 kg/10^3 g x 9.81 ms^{-2}

3.80 J to raise the psiton but total work will be p_{ex} DV + work to raise piston = 895 J + 3.80 J = **899 J**

2.6 (a) w = -p_{ex}DV
p_{ex} = 200 Torr x 133.3 Pa $Torr^{-1}$ = 2.666 x 10^4 Pa
DV = 3.3 L == 3.3 x 10^{-3} m^3
Therefore, w = -2.666 x 10^4 Pa x 3.3 x 10^{-3} m^3 = **-88 J**

(b) $w = -nRT \ln \frac{V_f}{V_i}$

$$n = \frac{4.50\,g}{16.04\,g mol^{-1}} = 0.2805 \text{ mol}$$

RT = 2.577 kJ mol^{-1}, V_i = 12.7 L, V_f = 16.0 L

$$w = -0.2805 \text{ mol} \times 2.577 \text{ kJ mol}^{-1} \times \ln\frac{16.0\text{L}}{12.7\text{L}} = \underline{\mathbf{-167\ J}}$$

2.7 $w = -nRT\ln\frac{V_f}{V_i}$

$V_f = \frac{1}{3}V_i$

$nRT = 52.0 \times 10^{-3} \text{ mol} \times 8.314 \text{ J K}^{-1} \text{ mol}^{-1} \times 260 \text{ K}$
$= 1.124 \times 10^2 \text{ J}$

$w = -\ 1.124 \times 10^2 \text{ J} \times \ln \frac{1}{3} = \underline{\mathbf{+123\ J}}$

2.8 $w = -p_{ex}DV$
$p_{ex} = 95.2 \text{ bar} = 95.2 \times 10^5 \text{ Pa}$

$DV = -0.550 \text{ L} \times 0.43 = -0.237 \text{ L} = -0.237 \times 10^{-3} \text{ m}^3$

Substituting in and keeping track of the minus sign

$w = (-95.2 \times 10^5 \text{ Pa}) \times (-0.237 \times 10^{-3} \text{ m}^3) = \underline{\mathbf{+2.25\ KJ}}$

2.9 $w = -p_{ex}DV$

$Mg(s) + 2HCl(aq) \rightarrow H_2(g) + MgCl_2(aq)$ $M(Mg) = 24.31 \text{ g mol}^{-1}$

$V_i = 0,\ V_f = \frac{nRT}{p_f},\ p_f = p_{ex}$

$w = -p_{ex}(V_f - V_i) = -p_{ex} \times \frac{nRT}{p_{ex}} = -nRT$

The number of moles of $H_2(g)$ produced are the same as the number of moles of Mg in the starting reaction.

$n = \frac{12.5\text{g}}{24.31\text{g mol}^{-1}} = 0.514 \text{ mol},\ RT = 2.438 \text{ kJ mol}^{-1}$

Substituting in to the above equation,
w = -0.514 mol x 2.438 kJ mol^{-1} = **-1.25 kJ**

2.10 $\Delta H^{\ominus}_{fus}$ = 2.60 kJ mol^{-1} [Table 2.2]

$$n = \frac{224 \times 10^3\ g}{22.99\ g\ mol^{-1}} = 2.74 \times 10^3\ mol$$

$q = n\Delta H^{\ominus}_{fus}$ = 2.74 x 10^3 x 2.60 kJ mol^{-1} = **2.53 x 10^4 kJ**

2.11 (a) at 25 °C DH = 44.0 kJ mol^{-1}

(b) at 100 °C DH = 40.7 kJ mol^{-1}

1.00 x 10^3 g x 1 mol/18.0 g = 55.5 mol

(a) at 25 °C

q = 44 kJ/mol x 55.5 mol = **2.4 x 10^3 kJ**

(b) 100 °C

q = 40.7 kJ/mol x 55.5 mol = **2.26 x 10^3 kJ**

2.12 q = CDT

$$C = \frac{q}{\Delta T} = \frac{124\ J}{5.23\ K} = \mathbf{23.7\ JK^{-1}}$$

2.13 C = 75.3 $JK^{-1}mol^{-1}$

250 g x 1 mol/18.0 g = 13.9 mol

$q = mC_mDT$ = 13.9 mol x 75.3 $Jmol^{-1}K^{-1}$ x 40 K

= **42 kJ**

2.14 First step is to melt 100 g of ice.

100 g x 1 mol/18.0 g = 5.55 mol

5.55 mol x DH_{fus} = 5.55 mol x 6.01 kJ mol^{-1} = 33.4 kJ

Next H_2O (l) form 0 °C to 100 °C
100 g x 4.18 $JK^{-1}g^{-1}$ x 100 K = 41.8 kJ

Vaporize H_2O

5.55 mol x 40.7 kJ mol^{-1} = 226 kJ

Summing all three steps:

33.4 kJ + 41.8 kJ + 226 kJ = **301 kJ**

2.15 $C = \frac{q}{\Delta T} = \frac{229\,J}{2.55\,K} = 89.9\ J\ K^{-1}$

The molar heat capacity at constant pressure is therefore

$C_p = \frac{89.8\,JK^{-1}}{3.0\,mol}$ = **30 J $K^{-1}mol^{-1}$**

For a perfect gas
$C_p - C_V = R$

Rearranging,
$C_V = C_p - R = (30 - 8.3)\ J\ K^{-1}\ mol^{-1}$ = **22 J K^{-1} mol^{-1}**

2.16 q = CDT, C = nC_p, V = 1072 m^3

$$n = \frac{pV}{RT} = \frac{1.0\,atm \times 1072 \times 10^3\,L}{(8.206 \times 10^{-2}\,L\,atm\,K^{-1}\,mol^{-1}) \times (298\,K)} = 4.38 \times 10^4\ mol$$

q = (4.38 x 10^4 mol) x (21 J K^{-1} mol^{-1}) x 10 K = **9.2 x 10^3 kJ**

Since q = P x t where P is the power of the heater and t is the time for which it operates,

$$t = \frac{q}{P} = \frac{9.2 \times 10^6\,\mathrm{J}}{1.5 \times 10^3\,\mathrm{J\,s^{-1}}} = \mathbf{\underline{6.1 \times 10^3\ s}}$$

In practice the walls and furniture of a room are also heated.

2.17 q = **<u>-1.2 kJ</u>** (energy leaves the sample)
Since DH = q at constant pressure
DH = **<u>-1.2 kJ</u>**

$$C = \frac{q}{\Delta T} = \frac{1.2\,\mathrm{kJ}}{15\,\mathrm{K}} = \mathbf{\underline{80\ J\ K^{-1}}}$$

2.18 q = CDT = nC_pDT
= 3.0 mol x (29.4 J K^{-1} mol^{-1}) x 25 K = +2.2 kJ

DH = q = **<u>+2.2 kJ</u>**

DU = DH - D(pV) = DH - D(nRT)
= DH - nRDT
= 2.2 kJ - 3.0 mol x 8.314 J K^{-1} mol^{-1} x 25 K
= 2.2 kJ - 0.62 kJ = **<u>+ 1.6 kJ</u>**

2.19 q = 1.50 mol x 26.0 kJ mol^{-1} = **<u>+ 39 kJ</u>**

$w = -p_{ex}DV \approx -p_{ex}V(g)$ since V(g)>>V(l)

$$\approx -p_{ex}\frac{nRT}{p_{ex}} = -nRT$$

Substituting into the above equation,

w $\approx$ -1.50 mol x 8.314 J K^{-1} mol^{-1} x 250 K = **<u>-3.12 kJ</u>**

DH = q = **<u>+39 kJ</u>**

DU = q + w = +39 kJ - 3.12 kJ = **+ 36 kJ**

2.20 a) 1.00 mol N_2

DH = 1.00 mol x (-92.22 kJ mol^{-1}) = **-92.22 kJ**

b) 1.00 mol NH_3 formed

$\frac{1 \text{ mol } N_2}{2 \text{ mol } NH_3}$ x 1 mol NH_3 x (-92.22 kJ mol^{-1}) = **-46.11 kJ**

2.21 a) Combustion of ethane is for one mole, therefore

1/2 (-3120 kJ mol^{-1}) = **1560 kJ**

b)3.00/4.00 (-3120 kJ mol^{-1}) = **2340 kJ**

2.22 $C_6H_5C_2H_5(l) + \frac{21}{2} O_2(g) \rightarrow 8CO_2\ (g) + 5H_2O(l)$

$\Delta H_c^\ominus = 8\Delta H_{fus}^\ominus\ (CO_2\ (g\ ) + 5\ \Delta H_{fus}^\ominus H_2O(l\ ) - \Delta H_{fus}^\ominus(\text{ethylbenzene}, l)$

$= 8\ (-393.51) + 5(-\ 285.83) - (\ -12.5)$ kJ mol^{-1}
$=$ **-4564.7 kJ mol^{-1}**

2.23 The reaction wanted is:

$C_6H_{10}\ (l) + H_2\ (g) \rightarrow C_6H_{12}\ (l\ )$

Use the following reactions:

1.) $C_6H_{10}\ (l) + 8\frac{1}{2} O_2 \rightarrow 6CO_2(g) + 5\ H_2O(l)$

$\Delta H_c^\ominus = -3752$ kJ mol^{-1}

2.) $C_6H_{12}\,(l\,) + 9O_2\,(g\,) \rightarrow 6\,CO_2(g\,) + 6\,H_2O(l\,)$
$\Delta H_c^{\ominus}$ = -3953 kJ mol^{-1}

3.) $H_2O(l\,) \rightarrow H_2(g\,) + \frac{1}{2}O_2(g\,)$ $\Delta H^{\ominus}$ = +286 kJ mol^{-1}

Summing eautions (1) and the reverse of equations (2) and (3) gives the desired equation.

$\Delta H^{\ominus}$ = -3752 kJ mol^{-1} + 3953 kJ mol^{-1} -286 kJ mol^{-1}
= **-85 kJ mol^{-1}**

2.24 $3C(s\,) + 3H_2(g\,) + O_2\,(g\,) \rightarrow CH_3COOCH_3(l\,)$
$\Delta H_f^{\ominus}$ = -442 kJ mol^{-1}

DU = DH - Dn$_g$RT, Dn$_g$ = -4 mol
Dn$_g$RT = -4 mol x 2.479 kJ mol^{-1} = -9.916 kJ

Therefore
$\Delta U_f^{\ominus}$ = -442 kJ mol^{-1} - (-9.916 kJ) = **-432 kJ mol^{-1}**

2.25 $C_{10}H_8(s) + 12O_2(g) \rightarrow 10\,CO_2(g\,) + 4H_2O(l\,)$
$\Delta H_c^{\ominus}$ = -5157 kJ mol^{-1}

The reverse reaction is
$10\,CO_2(g\,) + 4H_2O(l\,) \rightarrow C_{10}H_8(s) + 12O_2(g)$ $\Delta H^{\ominus}$ = 5157 kJ mol^{-1}

The CO_2 and H_2O can be replaced by adding the following two reactions and using

$\Delta H_f^{\ominus}(CO_2)$ and $\Delta H_f^{\ominus}$ (H_2O),[Table 2.10]

$10C(s\,) + 10O_2(g) \rightarrow 10\,CO_2(g\,)$ $\Delta H^{\ominus}$ = -3935 kJ mol^{-1}

$4H_2(g\,) + 2O_2(g) \rightarrow 4H_2O(l\,)$ $\Delta H^{\ominus}$ = -1143 kJ mol^{-1}

Overall

$$10C(s) + 4H_2(g) \rightarrow C_{10}H_8(s)$$

$\Delta H^{\ominus}$ = +5157 - 3935 - 1143 kJ mol^{-1} = **+79 kJ mol^{-1}**

2.26 $C = \dfrac{q}{\Delta T}$ and q = IVt

Therefore

$$C = \frac{3.20\,\text{A} \times 12.0\,\text{V} \times 27.0\,\text{s}}{1.617\,\text{K}} = \mathbf{641\ JK^{-1}}$$

Since 1 A V s = 1 J

2.27 $q = n\Delta H_c^{\ominus}$ and $\Delta H_c^{\ominus}$ = -5147 kJ mol^{-1} [Table 2.9]

Calculating q,

$$|q| = \frac{320 \times 10^{-3}\,\text{g}}{128.18\,\text{g mol}^{-1}} \times 5147\,\text{J mol}^{-1} = 12.85\ \text{kJ}$$

$$C = \frac{q}{\Delta T} = \frac{12.85\,\text{kJ}}{3.05\,\text{K}} = \mathbf{4.21\ kJ\ K^{-1}}$$

When phenol is used, $\Delta H_c^{\ominus}$ = -3054 kJ mol^{-1}

$$|q| = \frac{100 \times 10^{-3}\,\text{g}}{94.12\,\text{g mol}^{-1}} \times 3054\,\text{J mol}^{-1} = 3.245\ \text{kJ}$$

$$\Delta T = \frac{q}{C} = \frac{3.245\,\text{kJ}}{4.21\,\text{kJ K}^{-1}} = \mathbf{0.770\ K}$$

2.28 $q = C\Delta T$, $|\Delta H_c^\ominus| = \frac{q}{n} = \frac{C\Delta T}{n} = \frac{MC\Delta T}{m}$ where m is the mass of the sample.
Since M = 180.16 g mol^{-1},

$$|\Delta H_c^\ominus| = \frac{180.16\,g\,mol^{-1} \times 641\,J\,K^{-1} \times 7.793\,K}{0.3212\,g} = 2802\ kJ\ mol^{-1}$$

Since the combustion is exothermic, $\Delta H_c^\ominus$ = **-2.80 MJ mol^{-1}**

The combustion reaction is

$$C_6H_{12}O_6(s) + 6O_2(g) \rightarrow 6CO_2(g) + 6H_2O(l) \qquad Dn_g = 0$$

Therefore $DU_c = DH_c$, and DU_c = **-2.80 MJ mol^{-1}**

For the enthalpy of formation we combine the following equations:

$6CO_2(g) + 6H_2O(l) \rightarrow C_6H_{12}O_6(s) + 6O_2(g)$
DH = +2.8 MJ mol^{-1}

$6C(s) + 6O_2(g) \rightarrow 6CO_2(g)$
DH = -2.36 MJ mol^{-1}

$6H_2(g) + 3O_2(g) \rightarrow 6H_2O(l)$
DH = -1.72 MJ mol^{-1}

The sum of the three equations is

$$6C(s) + 6H_2(g) + 3O_2(g) \rightarrow C_6H_{12}O_6(s)$$

and DH_f = 2.80 - 2.36 - 1.72 MJ mol^{-1} = **-1.28 MJ mol^{-1}**

2.29 $AgBr(s) \rightarrow Ag^+(aq) + Br^-(aq)$

$\Delta H^\ominus = \Delta H_f^\ominus(Ag^+, aq) + \Delta H_f^\ominus(Br^-, aq) - \Delta H_f^\ominus(AgBr, (s)$

= 105.58 + (-121.55) - (-100.37) kJ mol^{-1}

= **+84.40 kJ mol^{-1}**

2.30 $NH_3SO_2 \rightarrow NH_3 + SO_2$ $\Delta H^{\ominus} = +40\ kJ\ mol^{-1}$

$NH_3 + SO_2 \rightarrow NH_3SO_2$ $\Delta H^{\ominus} = -40\ kJ\ mol^{-1}$

$\Delta H_f^{\ominus}\ (NH_3SO_2, s) = \Delta H_f^{\ominus}\ (NH_3, g) + \Delta H_f^{\ominus}(SO_2, g) - 40\ kJ\ mol^{-1}$

$= -46.11 - 296.83 - 40\ kJ\ mol^{-1} =$ **$\underline{-383\ kJ\ mol^{-1}}$**

2.31 $C(gr) + O_2\ (g) \rightarrow CO_2(g)$ $\Delta H^{\ominus} = -393.51\ kJ\ mol^{-1}$

$C(d) + O_2\ (g) \rightarrow CO_2(g)$ $\Delta H^{\ominus} = -395.41\ kJ\ mol^{-1}$

The difference in the two equations is

$C(gr) \rightarrow C(d)$ $\Delta H_{trs}^{\ominus} = -393.51 - (-395.41)\ kJ\ mol^{-1}$
$=$ **$\underline{+1.90\ kJ\ mol^{-1}}$**

2.32 $DU = -p_{ex}DV + q$

and $q = DH$

$p_{ex}DV = (-\ 150\ kPa) \times (-1.916 \times 10^{-3}\ L) = 0.2875\ J$

$1\ mol\ C = 12.01\ g \times 1\ cm^3/2.250\ g \times 1\ L/10^3\ cm^3$
$= 5.338 \times 10^{-3}\ L$

$1\ mol\ diamond = 12.01\ g \times 1\ cm^3/3.510\ g \times 1\ L/10^3\ cm^3$

$= 3.422 \times 10^{-3}\ L$

$V_D - V_C = 3.422 \times 10^{-3}\ L - 5.338 \times 10^{-3}\ L = -1.916 \times 10^{-3}\ L$

$DU = 0.2875\ kJ\ mol^{-1} + 1.90\ kJ\ mol^{-1} =$ **$\underline{2.19\ kJ\ mol^{-1}}$**

2.33 $q = n\Delta H_c^{\ominus}$

$$= \frac{1.5\,g}{342.3\,g\,mol} \times (-5645\ kJ\ mol^{-1}) = \underline{\mathbf{-25\ kJ}}$$

Effective work available is ≈ 25 kJ x 0.20 = 5.0 kJ
Since w = mgh, assume m ≈ 68 kg

$$h \approx \frac{5.0 \times 10^3\ J}{68\,kg \times 9.81\,m\,s^{-2}} = \underline{\mathbf{7.5\ m}}$$

2.34 $C_3H_8(l) \xrightarrow{\Delta H^\ominus_{vap}} C_3H_8(g)$

$C_3H_8(g) + 5\ O_2(g) \xrightarrow{\Delta H^\ominus_c} 3CO_2(g) + 4H_2O(l)$

$\Delta H^\ominus_c(l) = \Delta H^\ominus_{vap} + \Delta H^\ominus_c(g)$
$= 15\ kJ\ mol^{-1} - 2220\ kJ\ mol^{-1} = \underline{\mathbf{-2205\ kJ\ mol^{-1}}}$

(b) $Dn_g = -2$ [$5O_2$ replaced by $3CO_2$]
Therefore $\Delta U^\ominus_c(l) = \Delta H^\ominus_c(l) - (-2)RT$

$= -2205\ kJ\ mol^{-1} + 2 \times 2.479\ kJ\ mol^{-1}$
$= \underline{\mathbf{-2200\ kJ\ mol^{-1}}}$

2.35 a) exothermic, negative DH
b) endothermic
c) endothermic
d) endothermic
e) endothermic

2.36 (a) $\Delta H^\ominus = \Delta H^\theta_f(N_2O_4, g) - 2\Delta H^\theta_f(NO_2, g)$
$= 9.16 - 2 \times 33.18\ kJ\ mol^{-1} = \underline{\mathbf{-57.20\ kJ\ mol^{-1}}}$

(b) $\Delta H^\ominus = 1/2\Delta H^\theta_f(N_2O_4, g) - \Delta H^\theta_f(NO_2, g)$

$= 1/2(9.16) - 33.18\ \text{kJ mol}^{-1} =$ **$-28.6\ \text{kJ mol}^{-1}$**

(c) $\Delta H^\ominus = 2 \times \Delta H_f^\theta(\text{HNO}_3, aq) + \Delta H_f^\theta(\text{NO}, g)$
$- 3 \times \Delta H_f^\theta(\text{NO}_2, g) - \Delta H_f^\theta(\text{H}_2\text{O}, l)$

$= 2 \times (-207.36) + 90.25 - 3 \times (33.18) - (-285.83)\ \text{kJ mol}^{-1}$

= **$-138.2\ \text{kJ mol}^{-1}$**

(d) $\Delta H^\ominus = \Delta H_f^\ominus$ (propane, g) $- \Delta H_f^\ominus$ (cyclpropane, g)

$20.42 - 53.30\ \text{kJ mol}^{-1} =$ **$-32.88\ \text{kJ mol}^{-1}$**

(d) In order to calculate $\Delta H^\ominus$ first write the net ionic equation:

$H^+(ag) + Cl^-(aq) + Na^+(aq) + OH^-(aq) \rightarrow$
$Na^+(aq) + Cl^-(aq) + H_2O(l)$

Simplifying we obtain
$H^+(ag) + OH^-(aq) \rightarrow H_2O(l)$

$\Delta H^\ominus = \Delta H_f^\ominus(H_2O, l) - \Delta H_f^\ominus(H^+, aq) - \Delta H_f^\ominus(OH^-, aq)$
$= -285.83 - 0 - (-229.99)\ \text{kJ mol}^{-1} =$ **$-55.84\ \text{kJ mol}^{-1}$**

2.37 The formation of N_2O_5 is the sum of the three reaction

	$\Delta H^\ominus/(\text{kJ mol}^{-1})$
$2NO(g) + O_2(g) \rightarrow 2NO_2(g)$	-114.1
$\frac{1}{2}O_2(g) + 2NO_2(g) \rightarrow 2N_2O_5(g)$	$\frac{1}{2}(-110.2)$
$N_2(g) + O_2(g) \rightarrow 2NO(g)$	180.5
$N_2(g) + \frac{5}{2}O_2(g) \rightarrow N_2O_5(g)$	$+11.3$

Therefore $\Delta H_f^{\ominus}$ (N_2O_5 ,*g*) = **<u>+11.3 kJ mol⁻¹</u>**

2.38 $\Delta H^{\ominus}(T_2) = \Delta H^{\ominus}(T_1) + \Delta C_p \Delta T$

$\Delta C_p = C_p(N_2O_4, g) - 2C_p(NO_2, g)$

$= 77.28 - 2 \times 37.20\ \text{J K}^{-1}\ \text{mol}^{-1} = +2.88\ \text{J K}^{-1}\ \text{mol}^{-1}$

$\Delta H^{\ominus}(373\ \text{K}) = \Delta H^{\ominus}(298\ \text{K}) + \Delta C_p \Delta T$
$= -57.20\ \text{kJ mol}^{-1} + 2.88\ \text{J K}^{-1} \times 75\ \text{K}$
$= -57.20 + 0.22\ \text{kJ mol}^{-1} =$ **<u>$-56.98\ \text{kJ mol}^{-1}$</u>**

2.39 a) Change will be 2 x 4R - 3 x 7/2 R, **<u>decrease</u>** with increasing temperature

b) **<u>decrease</u>**, 8R - 7/2R - 3 x 7/2 R = -6R

c) **<u>increase</u>**, 8R + 7/2R - 4R - 2 x 7/2 R

2.40 a) 2 x 9R - 3 x 7/2 R **<u>increase</u>**

b) **<u>increase</u>**

2.41 **<u>Higher</u>**, because as temperature increases, DC_p is likely to increase because of the large n for products.

Additional problems

2.1 Calculate the work done when 1.0 mol O_2 that is confined in a balloon of volume 1.0 L at 25° C expands isothermally and reversibly to 5.0 L.

2.2 How much work is required to raise a 1.0 kg weight to a height of 2.0 m? How much work is required to move a 70 ton satellite 15 m into a shuttle cargo bay in an orbiting spacecraft ($g \approx 0$) ?

2.3 Calculate the work done to compress isothermally and reversibly 0.50 mol of perfect gas at 260 K from 5.0 L to 0.5 L.

2.4 The specific heat capacity of aluminum is 0.900 J K^{-1} g^{-1}.
(a) How much energy is required to raise the temperature of 10 g of aluminum from 20° C to 40° C?
(b) How much energy is required to raise the temperature of aluminum from 0° C to 20° C ?

2.5 Calculate the heat required to melt 250 kg of ice ay 0° C.

2.6 A 25 g sample of metal at 500 K is placed in a calorimeter containing 100.0 g water at 298 K. The final temperature of the water and metal is 405 K. Calculate the heat capacity of the metal assuming no heat escapes to the surroundings or is transferred to the calorimeter.

2.7 The standard enthalpy of combustion of methane is -890 kJ mol^{-1}. Calculate its standard enthalpy of formation.

2.8 Classify the following reactions as exothermic or endothermic
(a) KBr dissolved in water (the solution gets colder).
(b) Concentrated acid added to water (the solution gets hotter).
(c) Combustion of methane
(d) Fermentation of glucose ($\Delta H^{\ominus}$ = -67 kJ mol^{-1})

2.9 Calculate ΔH° for $2C\,(s) + O_2\,(g) \rightarrow 2CO\,(g)$

given ΔH_c° for solid C to form CO_2 is -393.7 kJ mol^{-1} and ΔH_c° for CO to form CO_2 is -283.3 kJ mol^{-1}.

2.10 A nutritional calorie is equal to 1000 cal (thermodynamic calories) = 1 kcal. A piece of apple pie contains 425 calories. How many 16 cm steps must a 120 lb (54.43 kg) woman climb to expend 425 nutritional calories?

2.11 Calculate the work done by a gas to expand reversilbly against a piston ($\Delta V = 2.0$ L) if the external pressure is a) 1.0 atm b) zero atm

2.12 Calculate ΔH° for the combustion of methanol.

$$2\,CH_3OH\,(l) + 3\,O_2\,(g) \rightarrow 2\,CO_2\,(g) + 4\,H_2O\,(l)$$

2.13 Using enthalpies of formation, calculate the standard enthalpy change for the reaction

$$2\,Al\,(s) + Fe_2O_3\,(s) \rightarrow Al_2O_3\,(s) + 2\,Fe\,(s)$$

$\Delta H_f^{\circ}\,(Fe_2O_3\,(s)) = -826$ kJ mol^{-1}

$\Delta H_f^{\circ}\,(Al_2O_3\,(s)) = -1676$ kJ mol^{-1}

2.14 When 2.0 mol O_2 is heated at a constant pressure of 3.0 atm, its temperature increases from 240 K to 270 K. The molar heat capacity of O_2 at constant pressure is 29.4 J K^{-1} mol^{-1}, calculate q, ΔH and ΔU .

Chapter 3

Thermodynamics: the Second Law

3.1 $\Delta S = \frac{q}{T} = \frac{120\ \text{J}}{293\ \text{K}}$ = **<u>0.41 JK^{-1}</u>**

3.2 a) $\Delta S = \frac{q}{T} = \frac{33\ \text{kJ}}{273\ \text{K}}$ = **<u>0.12 kJ K^{-1}</u>**

b) Water loses entropy

$\Delta S = \frac{-33\ \text{kJ}}{273\ \text{K}}$ = **<u>-0.12 kJ K^{-1}</u>**

3.3 $q = C_p\Delta T$ Calculate the amount of substance (in moles) to get the total heat extracted.

$$q = \frac{1.25 \text{x} 10^3\ \text{g}}{26.98\ \text{g mol}^{-1}} \text{ x } 24.35\ \text{J K}^{-1}\ \text{mol}^{-1} \text{x } 40\ \text{K}$$

= **<u>-45.1 kJ</u>**

$$\Delta S = C_p \ln \frac{T_2}{T_1}$$

$$= \frac{1.25 \text{x} 10^3\ \text{g}}{26.98\ \text{g mol}^{-1}} \text{x } 24.35\ \text{J K}^{-1}\ \text{mol}^{-1} \text{ x} \ln \frac{260\ \text{K}}{300\ \text{K}}$$

= **<u>-161 J K^{-1}</u>**

To treat the problem so that C_p is not constant with temperature use $C_p = a + bT$ [a and b available from tables] and integrate the expression with respect to T.

3.4 For the first step, melting 100 g ice:

$$\Delta S_{fus} = \frac{\Delta H_{fus}}{T_b} = \frac{6.01 \text{ kJ mol}^{-1}}{273 \text{ K}} \text{x } 100 \text{ g x} \frac{1 \text{ mol}}{18.0 \text{ g}}$$

= **<u>122 J K^{-1}</u>**

For the second step, heating the water:

$$\Delta S = C_V \ln\frac{T_f}{T_i} = 4.18 \text{ J K}^{-1}\text{g}^{-1} \text{ x } 100 \text{ g x } \ln \frac{373}{273}$$

= **<u>130 J K^{-1}</u>**

For the third step, vaporization:

$$\Delta S_{vap} = \frac{\Delta H_{vap}}{T_b} = \frac{40.7 \text{ kJ mol}^{-1}}{373 \text{ K}} \text{x } 100 \text{ g x} \frac{1 \text{ mol}}{18.0 \text{ g}}$$

= **<u>606 J K^{-1}</u>**

DS = 122 + 130 + 606 J K^{-1} = **<u>858 J K^{-1}</u>**

A graph of temperature vs time would show a constant 273 K temperature until all the ice had melted. Temperature would increase until the boiling point, 373 K was reached. Temperature would again remain constant until all the liquid was vaporized.

A graph of enthalpy or entropy vs time would show an increase in the solid to liquid phase, a steeper slope in the liquid heating time and, for entropy a very sharp increase as gas is formed.

3.5 $\Delta S = nR\ln\frac{V_f}{V_i}$ Substituting in for V = nRT/p

$$= nR\ln\frac{p_f}{p_i}$$

$$= \frac{25\,g}{16.04\,g\,mol^{-1}} \times 8.314\,J\,K^{-1}\,mol^{-1} \times \ln\frac{185}{2.5} = \underline{\mathbf{56\ J\ K^{-1}}}$$

3.6 $\Delta S = nR\ln\frac{V_f}{V_i}$ use the perfect gas law to calculate nR

$nR = \frac{p_i V_i}{T_i} = \frac{1.00\,atm \times 15.0\,L}{250\,K}$ converting units

$$= \frac{1.013 \times 10^5\,Pa \times 15.0 \times 10^{-3}\,m^3}{250\,K} = 6.08\ J\ K^{-1}$$

$$\ln\frac{V_f}{V_i} = \frac{\Delta S}{nR} = \frac{-10.0\,J\,K^{-1}}{6.08\,J\,K^{-1}} = -1.64$$

Therefore, V_i = 15.0 L and

V_f = **2.90 L**

3.7 Calculate in two steps:

DS for compression

$$DS = 8.314\ JK^{-1}mol^{-1} \times 1\ mol \times \ln\frac{0.500\ L}{2.0\ L} = -11.5\ J$$

Second step heat from 300 K to 400 K

$$\Delta S = C_V \ln\frac{T_f}{T_i} = \frac{3}{2}\left(8.314\ J\ K^{-1}\right)\ln\frac{400}{300} = 3.59\ J$$

Total DS = - 11.5 J + 3.59 J = **-7.9 J**

3.8 $V_f = 2V_i$

$$\Delta S = nR\ln\frac{V_f}{V_i} = nR\ln 2$$

$$\Delta S = C_V \ln\frac{T_f}{T_i} = nR\ln 2 = -3/2R\ln\frac{T_f}{T_i}$$

$$\ln\frac{T_f}{T_i} = -0.4621$$

T_f = **$0.630\ T_i$**

3.9 Step 1-> 2 isothermal, $\Delta S = nR\ln\frac{V_f}{V_i}$

2-> 3 adiabatic, DS = 0, 3 -> 4 isothermal compression,
$\Delta S = nR\ln\frac{V_f}{V_i}$ where $V_f = V_i$ in step one.

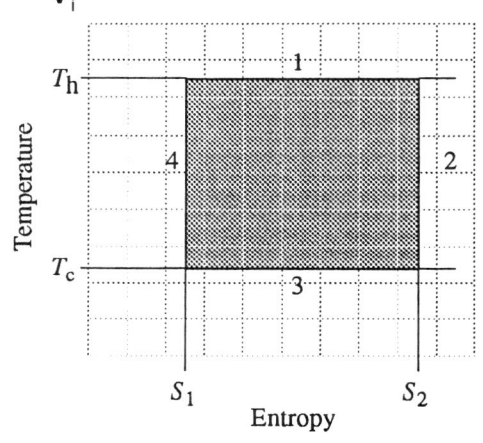

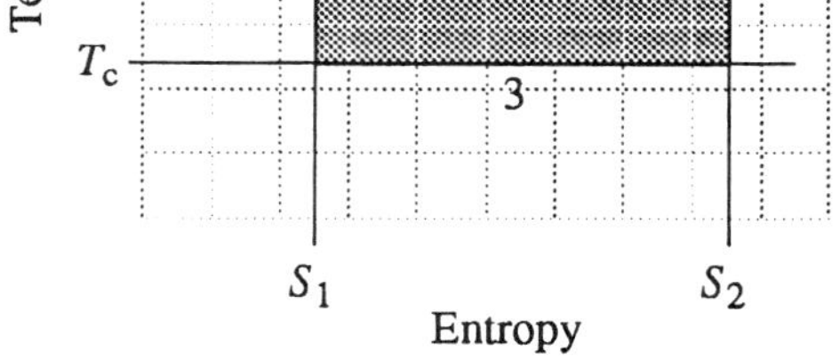

3.10 First find the common final temperature, T_f, by noting that the heat lost by the hot sample is gained by the cold sample.

$n_iC_p(T_f - T_{i1}) = n_2C_p(T_f - T_{i2})$

Solving for T_f

$$T_f = \frac{n_1T_{i1} + n_2T_{i2}}{n_1 + n_2}$$

Since $n_1 = n_2$
$T_f = 1/2(353\ K + 283\ K) = 318\ K$

The total entropy change is therefore

$$\Delta S = \Delta S_1 + \Delta S_2 = n_1C_p \ln\frac{T_f}{T_{i1}} + n_2C_p \ln\frac{T_f}{T_{i2}}$$

$$= 55.5\ \text{mol} \times 75.5\ \text{J K}^{-1}\ \text{mol}^{-1} \times \left(\ln\frac{318}{353} + \ln\frac{318}{283}\right)$$

= **$\underline{5.3\ \text{J K}^{-1}}$**

3.11 $\Delta S_{vap} = \frac{\Delta H_{vap}}{T_b} = \frac{29.4 \times 10^3\ \text{J mol}^{-1}}{334.88\ \text{K}} =$ **$\underline{+87.8\ \text{J K}^{-1}\ \text{mol}^{-1}}$**

since vaporization occurs reversilby, $\Delta S_{vap} = 0$ so
$\Delta S_{surr} =$ **$\underline{-87.8\ \text{J K}^{-1}\ \text{mol}^{-1}}$**

3.12 (a) $\Delta S^{\ominus} = 2S^{\ominus}(CH_3COOH, l) - 2S^{\ominus}(CH_3CHO, g)$
$- S^{\ominus}(O_2, g)$

$= 2 \times 159.8 - 2 \times 250.3 - 205.14\ \text{J K}^{-1}\ \text{mol}^{-1}$
= **$\underline{-386.1\ \text{J K}^{-1}\ \text{mol}^{-1}}$**

(b) $\Delta S^{\ominus} = 2S^{\ominus}(AgBr, s) + S^{\ominus}(Cl_2, g) - 2S^{\ominus}(AgCl, s)$
$- S^{\ominus}(Br_2, l)$

$= 2 \times 107.1 + 223.07 - 2 \times 96.2 - 152.23\ \text{J K}^{-1}\ \text{mol}^{-1}$
= **$\underline{+92.6\ \text{J K}^{-1}\ \text{mol}^{-1}}$**

(c) $\Delta S^\ominus = S^\ominus(HgCl_2, s) - S^\ominus(Hg, l) - S^\ominus(Cl_2, g)$

$= 146.0 - 76.02 - 223.07\ \mathrm{J\ K^{-1}\ mol^{-1}}$
$= \underline{\mathbf{-153.1\ J\ K^{-1}\ mol^{-1}}}$

(d) $\Delta S^\ominus = S^\ominus(Zn^{2+}, aq) + S^\ominus(Cu, s) - S^\ominus(Zn, s) - S^\ominus(Cu^{2+}, aq)$

$= -112.1 + 33.15 - 41.63 + 99.6\ \mathrm{J\ K^{-1}\ mol^{-1}}$
$= \underline{\mathbf{-21.0\ J\ K^{-1}\ mol^{-1}}}$

(e) $\Delta S^\ominus = 12\,S^\ominus(CO_2, g) + 11\,S^\ominus(H_2O, l) - S^\ominus(C_{12}H_{22}O_{11}, s) - 12\,S^\ominus(O_2, g)$

$= (12 \times 213.74) + (11 \times 69.91) - 360.2 - (12 \times 205.14\ \mathrm{J\ K^{-1}\ mol^{-1}})$

$= \underline{\mathbf{+512.0\ J\ K^{-1}\ mol^{-1}}}$

3.13 $C_p = 7/2\ R = C_v + R$ for nonlinear $C_v = 5R$
$C_p = 4\ R = C_v + R$ for linear and $C_v = 3R$

$$\Delta S = C_v \ln \frac{T_f}{T_i}$$

a) $DS_{prod} - DS_{react} = -4R \ln \frac{283}{273} = \underline{\mathbf{1.20\ JK^{-1}}}$

b) $DS_{prod} - DS_{react} = -3/2R \ln \frac{283}{273} = \underline{\mathbf{0.449\ JK^{-1}}}$

3.14 $\Delta H^\ominus = -2808\ \mathrm{kJ\ mol^{-1}}$

No work done, so $DS = \dfrac{q_{rev}}{T} = \dfrac{-2808\ \mathrm{kJ\ mol^{-1}} \times 100\ \mathrm{g} \times \dfrac{1\ \mathrm{mol}}{180\ \mathrm{g}}}{273\ \mathrm{K} + 37\ \mathrm{K}}$

$= \underline{\mathbf{-5.03\ kJ\ K^{-1}}}$

3.15 $\Delta G^\ominus = \Delta H^\ominus - T\Delta S$

(a) $\Delta H^\ominus = 2\Delta H_f^\ominus(CH_3COOH, l) - 2\Delta H_f^\ominus(CH_3CHO, g)$

$= 2 \times (-484.5) - 2 \times (-166.19)\ \text{kJ mol}^{-1} = +53.40\ \text{kJ mol}^{-1}$

$\Delta G^\ominus = -636.62\ \text{kJ mol}^{-1} - 298.15\ \text{K} \times (-386.1\ \text{J K}^{-1}\ \text{mol}^{-1})$
$= \underline{\mathbf{-521.5\ \ kJ\ mol^{-1}}}$

(b) $\Delta H^\ominus = 2\Delta H_f^\ominus(AgBr, s) - 2\Delta H_f^\ominus(AgCl, s)$

$= 2 \times (-100.37) - 2 \times (-127.07)\ \text{kJ mol}^{-1}$
$= +53.40\ \text{kJ mol}^{-1}$

$\Delta G^\ominus = +53.40\ \text{kJ mol}^{-1} - 298.15\ \text{K} \times 92.6\ \text{J K}^{-1}\ \text{mol}^{-1}$

$= \underline{\mathbf{+25.8\ \ kJ\ mol^{-1}}}$

(c) $\Delta H^\ominus = \Delta H_f^\ominus(HgCl_2, s) = -224.3\ \text{kJ mol}^{-1}$

$\Delta G^\ominus = -224.3\ \text{kJ mol}^{-1} - 298.15\ \text{K} \times (-153.1\ \text{J K}^{-1}\ \text{mol}^{-1})$
$= \underline{\mathbf{-178.7\ kJ\ mol^{-1}}}$

(d) $\Delta H^\ominus = \Delta H_f^\ominus(Zn^{+2}, aq) - \Delta H_f^\ominus(Cu^{+2}, aq)$

$= +153.89 - 64.77\ \text{kJ mol}^{-1} = -218.66\ \text{kJ mol}^{-1}$

$\Delta G^\ominus = -218.66\ \text{kJ mol}^{-1} - 298.15\ \text{K} \times (-21.0\ \text{J K}^{-1}\ \text{mol}^{-1})$
$= \underline{\mathbf{-212.40\ kJ\ mol^{-1}}}$

(e) $\Delta H^\ominus = \Delta H_c^\theta = -5645\ \text{kJ mol}^{-1}$

$\Delta G^\ominus = -5645\ \text{kJ mol}^{-1} - 298.15\ \text{K} \times 512.0\ \text{J K}^{-1}\ \text{mol}^{-1}$

$= \underline{\mathbf{-5798\ kJ\ mol^{-1}}}$

3.16 (a) $\Delta G^\ominus = 2\Delta G_f^\ominus(CH_3COOH, l) - 2\Delta G_f^\ominus(CH_3CHO, g)$
$= 2 \times (-389.9) - 2 \times (-128.86)$ kJ mol^{-1}
= **-522.1 kJ mol^{-1}**

(b) $\Delta G^\ominus = 2\Delta G_f^\ominus(AgBr, s) - 2\Delta G_f^\ominus(AgCl, s)$

$= 2 \times (-96.90) - 2 \times (-109.79)$ kJ mol^{-1}
= **+25.78 kJ mol^{-1}**

(c) $\Delta G^\ominus = \Delta G_f^\ominus(HgCl_2, s)$ = **-178.6 kJ mol^{-1}**

(d) $\Delta G^\ominus = \Delta G_f^\ominus(Zn^{2+}, aq) - \Delta G_f^\ominus(Cu^{2+}, aq)$
= -147.06 - 65.49 kJ mol^{-1} = **-212.55 kJ mol^{-1}**

(e) $\Delta G^\ominus = 12\Delta G_f^\ominus(CO_2, g) + 11\Delta G_f^\ominus(H_2O, l) - \Delta G_f^\ominus(C_{12}H_{22}O_{11}, s)$

$= 12 \times (-394.36) + 11 \times (-237.13) - (-1543)$ kJ mol^{-1}

= **-5798 kJ mol^{-1}**

3.17 $\Delta H_c^\theta = -890$ kJ mol^{-1}

$$1.0 \times 10^3 \text{ g} \times \frac{1 \text{ mol}}{16.04 \text{ g}} = 62.3 \text{ mol}$$

heat = -890 kJ mol^{-1} x 62.3 mol = **5.5 x 10^4 kJ**

b) non-expansion work

DG = DH - TDS

DS = $2S^\ominus(H_2O, l) + S^\ominus(CO_2, g) - 2S^\ominus(O_2, g) - S^\ominus(CH_4, g)$

= 2 x 69.91 JK^{-1}mol^{-1} + 213.74 JK^{-1}mol^{-1}
- 2 x 205.14 JK^{-1}mol^{-1} - 186.26 JK^{-1}mol^{-1}

$= -242.98\ JK^{-1}mol^{-1}$

$DG = -890\ kJ\ mol^{-1} - (298\ K)(-242.98\ JK^{-1}mol^{-1})$
$= -817.6\ kJ\ mol^{-1}$

DG = $-817.6\ kJ\ mol^{-1} \times 62.3\ mol$ = **-5.1×10^4 kJ or 5.1×10^4 kJ of non-expansion work.**

3.18 a) heat

$\Delta H_c^{\theta} = -2.544\ kJ\ mol^{-1}$

$$1.0 \times 10^3\ g \times \frac{1\ mol}{180.16\ g} = 5.55\ mol$$

heat = $-2.544\ kJ\ mol^{-1} \times 5.55\ mol$ = **-1.4×10^4 kJ**

b) non-expansion work

DG = DH - TDS

$DS = 6S^{\ominus}(H_2O, g) + 6S^{\ominus}(CO_2, g) - 6S^{\ominus}(O_2, g) - S^{\ominus}(C_6H_{12},O_6\ g)$

$= 6 \times 188.83\ JK^{-1}mol^{-1} + 6 \times 213.74\ JK^{-1}mol^{-1} - 6 \times 205.14\ JK^{-1}mol^{-1} - 212\ JK^{-1}mol^{-1}$
$= 972\ JK^{-1}mol^{-1}$

$DG = -2544\ kJ\ mol^{-1} - (298\ K)(-0.972\ kJK^{-1}mol^{-1})$
$= -2.83 \times 10^3\ kJ\ mol^{-1}$

DG = $-2.83 \times 10^3\ kJ\ mol^{-1} \times 5.55\ mol$ = **-1.57×10^4 kJ or 1.57×10^4 kJ of non-expansion work.**

3.19 a) non-expansion work

water vapor

DG = DH - TDS

$= -5162 \text{ kJ mol}^{-1} - (298)(1.82 \text{ kJ mol}^{-1})$
$= -5.70 \times 10^3 \text{ kJ mol}^{-1}$

work = 2.92 mol x 5.73 x 10^3 kJ mol^{-1} = **1.67×10^4 kJ**

liquid water

DG = DH - TDS

$= -5645 \text{ kJ mol}^{-1} - 298 (0.512 \text{ kJ mol}^{-1})$
$= -5.80 \times 10^3 \text{ kJ mol}^{-1}$

work = 2.92 mol x 5.80 x 10^3 kJ mol^{-1} = **1.69×10^4 kJ**

Expansion work, water vapor

$w = -p_{ex}DV$

$DV = RT/P(n_2-n_1)$

w = -RT(Dn) = -2.4790 kJ mol^{-1} x 31.9 mol

= **-79.0 kJ expansion work**

Total work = -1.67 x 10^4 -79.0 kJ = **1.68×10^4 kJ**

If water is generated as a liquid then there is no significant expansion work and total work = DG.

3.20 To calculate the standard free energy of formation, we need to calculate the entropy of formation and the enthalpy of formation.

$$6C(s) + 3H_2(g) + \frac{1}{2}O_2(g) \rightarrow C_6H_5OH(s)$$

$$\Delta S^\ominus = S^\ominus C_6H_5OH(s) - 6S^\ominus(C, s) - 3S^\ominus(H_2, g) - \frac{1}{2}S^\ominus(O_2, g)$$

$= 144.0 - 6 \times 5.740 - 3 \times 130.68 - \frac{1}{2} \times 205.14$

$= -385.05$ kJ mol^{-1}

$C_6H_5OH(s) + 7O_2(g)) \rightarrow 6CO_2(g) + 3H_2O(l)$

$\Delta H_c^\theta = 6\Delta H_f^\ominus\ CO_2(g) + 3\ \Delta H_f^\ominus H_2O(l) - \ \Delta H_f^\ominus C_6H_5OH(s)$

Rearranging

$\Delta H_f^\ominus C_6H_5OH(s) = 6\,\Delta H_f^\ominus\ CO_2(g) + 3\ \Delta H_f^\ominus H_2O(l) - \Delta H_c^\theta$

$= 6 \times (-393.51) + 3(-285.83) - (-3054)$ kJ mol^{-1}
$= -164.55$ kJ mol^{-1}

Therefore

$\Delta G_f^\ominus = -164.55$ kJ mol^{-1} $- 298.15$ K $\times$ (-385.05 J K^{-1} mol^{-1})
= **-49.8 kJ mol^{-1}**

3.21 $CH_4(g) + 2O_2(g) \rightarrow CO_2(g) + 2H_2O(l)$
$\Delta G^\ominus = -817.90$ kJ mol^{-1}

Therefore, the maximum non-expansion work is **817.90 kJ mol^{-1}** since

$|w_e| = |\Delta G|$

3.22 $\Delta S_{trs} = \dfrac{\Delta H_{trs}}{T_{trs}} = \dfrac{+1.9\,\mathrm{kJ mol^{-1}}}{2000\mathrm{K}}$ = **+0.95 J K^{-1} mol^{-1}**

3.23 $\Delta G^\ominus = -RT \ln K$

$= -8.314$ J K^{-1} mol^{-1} $\times$ 400 K $\times \ln 2.07$

$= \underline{\mathbf{-2.42\ kJ\ mol^{-1}}}$

3.24 $K = e^{-\Delta G^{\ominus}/RT}$

$= e^{+3.67\times10^{3}\,\mathrm{J\,mol^{-1}}/8.314\,\mathrm{J\,K^{-1}\,mol^{-1}}\times 400\,\mathrm{K}} = \underline{\mathbf{3.01}}$

3.25 $\Delta G^{\ominus} = -RT\ln K$

$$\frac{\Delta G_1^{\ominus}}{\Delta G_2^{\ominus}} = \frac{-200\ \mathrm{kJ\ mol^{-1}}}{-100\ \mathrm{kJ\ mol^{-1}}} = \frac{-RT\ln K_1}{-RT\ln K_2}$$

$2\ln K_2 = \ln K_1$

$\underline{\mathbf{K_2^2 = K_1}}$

3.26 $K_1 = 10 \times K_2$

$\Delta G_2^{\ominus} = -RT\ln(0.10)K_1 = -RT\ln K_1 - RT\ln(0.10)$

$= -300\ \mathrm{kJ\ mol^{-1}} - (2.4790\ \mathrm{kJ\ mol^{-1}})(-2.303)$

$= \underline{\mathbf{-294\ kJ\ mol^{-1}}}$

3.27 $\ln K_{eq} = 0$

$\underline{\mathbf{K_{eq} = 1.0}}$

3.28 $$\frac{\Delta G^{\ominus\prime}}{T'} - \frac{\Delta G^{\ominus}}{T} = \Delta H^{\ominus}\left(\frac{1}{T'} - \frac{1}{T}\right)$$

and

$$\ln K' - \ln K = -\frac{\Delta H^{\ominus}}{R}\left(\frac{1}{T'} - \frac{1}{T}\right)$$

K' = 1 implies ln K' = 0 and $\Delta G^{\ominus}{}' = 0$, which occurs when

$$-\frac{\Delta G^{\ominus}}{T} = \frac{\Delta H^{\ominus}}{R}\left(\frac{1}{T'} - \frac{1}{T}\right)$$

$$\text{Or} \quad \frac{1}{T'} = \frac{1}{T} - \frac{\Delta G^{\ominus}}{\Delta H^{\ominus} T} = \frac{1}{T}\left(1 - \frac{\Delta G^{\ominus}}{\Delta H^{\ominus}}\right)$$

$$= \frac{1}{1280\,\text{K}}\left(1 - \frac{33\,\text{kJmol}^{-1}}{224\,\text{kJmol}^{-1}}\right) = 6.66 \times 10^{-4}\ \text{K}^{-1}$$

T' = **<u>1500 K</u>**

3.29 $\Delta G = \Delta G_f^{\circ}(CH_3CHO) + \Delta G_f^{\circ}(CO_2) - \Delta G_f^{\circ}(CH_3COCO_2^-)$

$= (-133 - 394.36 + 474)\ \text{kJ mol}^{-1} =$ **<u>-53.4 kJ mol^{-1}</u>**

3.30 We would expect ΔG to be more negative for a gaseous reaction because DS would be larger and positive.

3.31 Since ΔG values don't depend on the presence of a catalyst, ΔG values should be the same.

3.32 First substitute in and solve for K at 390 K and 410 K.

$$\ln K = -1.04 - \frac{1088\,\text{K}}{390} + \frac{1.51 \times 10^5\,\text{K}^2}{(390)^2}$$

ln K = - 2.83

Solving for K' where T = 410 K,

ln K' = - 2.80

Using the equation

$$\ln K' - \ln K = -\frac{\Delta H}{R}\left(\frac{1}{T'} - \frac{1}{T}\right)$$

Substitution gives:

$$0.041 = -\frac{\Delta H^{\ominus}}{8.314\,\mathrm{JK^{-1}mol^{-1}}}\left(\frac{1}{410} - \frac{1}{390}\right)$$

Solving for $\Delta H^{\ominus}$ gives

$\Delta H^{\ominus}$ = **<u>2.73 kJ mol^{-1}</u>**

$\Delta G^{\ominus}$ = -RTln K

$$= RT\,x\left(1.04 + \frac{1088}{T} - \frac{1.51x10^5}{T^2}\right)$$

$$= RT\,x\left(1.04 + \frac{1088}{400} - \frac{1.51x10^5}{(400)^2}\right) = +9.37\ \mathrm{kJ\ mol^{-1}}$$

$$= \Delta H^{\ominus} - T\Delta S^{\ominus}$$

Therefore,

$$\Delta S^{\ominus} = \frac{\Delta H^{\ominus} - \Delta G^{\ominus}}{T} = \frac{2.77\,\mathrm{kJmol^{-1}} - 9.37\,\mathrm{kJmol^{-1}}}{400\mathrm{K}}$$

= **<u>-16.5 J K^{-1} mol^{-1}</u>**

3.33 Since Q can be related to the partial pressures, first calculate the partial pressures.

$p_B = x_B p$ [B denotes borneol]

$$= \frac{0.15\,\mathrm{mol}}{0.15\,\mathrm{mol} + 0.30\,\mathrm{mol}}\ \mathrm{x\ 600\ Torr = 200\ Torr}$$

$p_I = p - p_B$ [I denotes isoborneol] $= 400$ Torr

$$Q = \frac{p_I}{p_B} = 2.00$$

$$\Delta G_f = \Delta G^{\ominus} + RT\ln Q$$

$= +9.4$ kJ mol^{-1} + 8.314 J K^{-1} mol^{-1} x 503 K x ln2.00
= **+12.3 kJ mol^{-1}**

3.34 $K = \frac{p_I}{p_B} = \frac{x_I}{x_b}$ $[p_I = x_I p, p_B = x_B p]$

$$= \frac{1 - x_B}{x_B}$$

Rearranging: $x_B = \frac{1}{1+K} = \frac{1}{1+0.106} = 0.904$

$x_I = 0.096$

The initial amounts of the isomers are

$$n_B = \frac{7.50\,g}{M},\ n_I = \frac{14.0\,g}{M},\ n = \frac{21.50\,g}{M}$$

The total amount remains the same, but at equilibrium

$\frac{n_B}{n} = x_B =$ **0.90**, and $x_I =$ **0.10**

3.35 $U(s) + \frac{3}{2}H_2(g) \leftrightarrow UH_3(s)$

$$\ln K = \ln\left(\frac{p}{p^{\ominus}}\right)^{-3/2} = -\frac{3}{2}\ln\left(\frac{p}{p^{\ominus}}\right)$$

$$\Delta G^{\ominus} = -RT \ln K = \frac{3}{2}\ln\left(\frac{p}{p^{\ominus}}\right)$$

$$= \frac{3}{2} \times 8.314\,\mathrm{JK^{-1}mol^{-1}} \times 500\,\mathrm{K} \times \ln\frac{1.04\,\mathrm{Torr}}{750\,\mathrm{Torr}}$$

$[p^{\ominus} = 1\text{ bar} = 1\text{ atm}]$

= **$-41.0\ \mathrm{kJ\ mol^{-1}}$**

3.36 Determine whether $\Delta G^{\ominus} < 0$ at 298 K;

(a) $\Delta G^{\ominus}/(\mathrm{kJ\ mol^{-1}}) = -202.87 - (-95.30 - 16.45) = -91.12$
(b) $\Delta G^{\ominus}/(\mathrm{kJ\ mol^{-1}}) = 3 \times (-856.64) - 2 \times (-1582.3) = +594.7$
(c) $\Delta G^{\ominus}/(\mathrm{kJ\ mol^{-1}}) = -100.4 - (-33.56) = -66.8$
(d) $\Delta G^{\ominus}/(\mathrm{kJ\ mol^{-1}}) = 2 \times (-33.56) - (-166.9) = +99.8$
(e) $\Delta G^{\ominus}/(\mathrm{kJ\ mol^{-1}}) = -744.53 - (-27.83) - 2 \times (-120.35) = -476.00$

Therefore **(a), (c) and (e)** have K > 1 at 298.

3.37 Determne whether $\Delta H^{\ominus} > 0$ at 298 K using $\Delta H_f^{\ominus}$ values.

(a) $\Delta H^{\ominus}/(\mathrm{kJ\ mol^{-1}}) = -314.43 - (-46.11 - 92.31) = -176.01$

(b) $\Delta H^{\ominus}/(\mathrm{kJ\ mol^{-1}}) = 3 \times (-910.94) - 2 \times (-1675.7) = +618.6$

(c) $\Delta H^{\ominus}/(\mathrm{kJ\ mol^{-1}}) = -100.0 - (-20.63) = -79.4$

(d) $\Delta H^{\ominus}/(\mathrm{kJ\ mol^{-1}}) = 2 \times (-20.63) - (-178.2) = +136.9$

(e) $\Delta H^{\ominus}/(\mathrm{kJ\ mol^{-1}}) = -909.27 - (-39.7) - 2 \times (-187.78) = -494.0$

Since (a), (c) and (e) are exothermic, an increase in temperature favors the reactants; **<u>(b) and (d)</u>** are endothermic, and an increase in temperature favors the products.

3.38 $\ln\frac{K'}{K} = \frac{\Delta H^{\theta}}{R}\left(\frac{1}{T} - \frac{1}{T'}\right)$

Solving for $\Delta H^{\ominus}$

$$\Delta H^{\ominus} = \frac{R \ln K'/K}{\frac{1}{T} - \frac{1}{T'}}$$

T' = 308 K, hence with K'/K = k

$$\Delta H^{\ominus} = \frac{8.314\,\mathrm{J\,K^{-1}\,mol^{-1}} \times \ln \kappa}{\left(\frac{1}{298\,\mathrm{K}} - \frac{1}{308\,\mathrm{K}}\right)} = 76\ \mathrm{kJ\ mol^{-1}} \times \ln k$$

Therefore

(a) k = 2, $\Delta H^{\ominus}$ = 76 kJ mol^{-1} x ln 2 = **<u>+53 kJ mol^{-1}</u>**

(b) k = 1/2, $\Delta H^{\ominus}$ = 76 kJ mol^{-1} x ln 1/2 = **<u>-53 kJ mol^{-1}</u>**

3.39 $\Delta G_f = \Delta G^{\ominus} + RT \ln Q$ for $\frac{1}{2}N_2(g) + \frac{3}{2}H_2(g) \rightarrow NH_3(g)$

$$Q = \frac{p(NH_3)/p^{\ominus}}{\left(p(N_2)/p^{\ominus}\right)^{1/2}\left(p(H_2)/p^{\ominus}\right)^{3/2}} = \frac{p(NH_3)}{p(N_2)^{1/2}p(H_2)^{3/2}}$$

$$= \frac{4.0}{(3.0)^{1/2}(1.0)^{3/2}} = \frac{4.0}{\sqrt{3.0}}$$

Therefore

$$\Delta G_f = -16.45\ \mathrm{kJ\ mol^{-1}} + RT \ln \frac{4.0}{\sqrt{3.0}}$$

$= -16.45 \text{ kJ mol}^{-1} + 2.07 \text{ kJ mol}^{-1} = \underline{\mathbf{-14.38 \text{ kJ mol}^{-1}}}$

Since $\Delta G_f < 0$, the spontaneous diretion of reaction is toward products.

3.40 DG = -16.5 kJ mol^{-1}

$$100 \text{ g } N_2 \times \frac{1 \text{ mol}}{28.02 \text{ g}} \times \frac{2 \text{ mol } NH_3}{1 \text{ mol } N_2} = 7.14 \text{ mol}$$

$w = (-16.5 \text{ kJ mol}^{-1}) \times 7.14 \text{ mol} = \underline{\mathbf{-118 \text{ kJ}}}$

3.41 $NH_4Cl(s) \rightleftharpoons NH_3(g) + HCl(g)$

$p = p(NH_3) + p(HCl) = 2p(NH_3) \quad [p(NH_3) = p(HCl)]$

$$\text{(a)} \quad K_p = \frac{p(NH_3)}{p^{\ominus}} \times \frac{p(HCl)}{p^{\ominus}} = \frac{p(NH_3)^2}{p^{\ominus 2}} = \frac{1}{4}\left(\frac{p}{p^{\ominus}}\right)^2$$

$$\text{At } 427^{\circ}\text{ C (700 K), Kp} = \frac{1}{4}\left(\frac{608\,\text{kPa}}{100\,\text{kPa}}\right)^2 = \underline{\mathbf{9.24}}$$

$$\text{At } 459^{\circ}\text{ C (732 K), } K_p = \frac{1}{4}\left(\frac{1115\,\text{kPa}}{100\,\text{kPa}}\right)^2 = \underline{\mathbf{31.08}}$$

(b)
$$\Delta G^{\ominus} = -RT \ln K_p$$
$$= -8.314 \text{ J K}^{-1} \text{ mol}^{-1} \times 700 \text{ K} \times \ln 9.24$$
$$= \underline{\mathbf{-12.9 \text{ kJ mol}^{-1}}} \text{ (at } 427^{\circ}\text{ C)}$$

$$\text{(c)} \quad \Delta H^{\ominus} \approx \frac{R \ln K'/K}{\left(\frac{1}{T} - \frac{1}{T'}\right)}$$

$$\approx \frac{8.314\,\text{JK}^{-1}\text{mol}^{-1} \times \ln \frac{31.08}{9.24}}{\left(\frac{1}{700\text{K}} - \frac{1}{732\text{K}}\right)} = \underline{\mathbf{+161 \text{ kJ mol}^{-1}}}$$

(d) $\Delta S^{\ominus} = \dfrac{\Delta H^{\ominus} - \Delta G^{\ominus}}{T} = \dfrac{161\,\text{kJ mol}^{-1} - (-12.9\,\text{kJmol}^{-1})}{700\,\text{K}}$

$= \underline{\mathbf{+248\ J\ K^{-1} mol^{-1}}}$

3.42 $\Delta G^{\ominus} = \Delta H^{\ominus} - T\Delta S^{\ominus} = 0$ when $\Delta H^{\ominus} = T\Delta S^{\ominus}$

Therefore the decomposition temperature (when K = 1) is

$$T = \frac{\Delta H^{\ominus}}{\Delta S^{\ominus}}$$

(a) $CaCO_3(s) \rightarrow CaO(s) + CO_2(g)$

$\Delta H^{\ominus}$ = -635.09 - 393.51 -(-1206.9) kJ mol^{-1} = +178.3 kJ mol^{-1}

$\Delta S^{\ominus}$ = 39.75 + 213.74 - 92.9 J K^{-1} mol^{-1} = +160.6 J K^{-1} mol^{-1}

$T = \dfrac{178.3 \times 10^{3}\ \text{Jmol}^{-1}}{160.6\ \text{JK}^{-1}\text{mol}^{-1}}$ = **<u>1110 K</u>** (840° C)

(b) $CuSO_4 \cdot 5H_2O(s) \leftrightarrow CuSO_4(s) + 5H_2O(g)$

$\Delta H^{\ominus}$ = -771.36 + 5 x (-241.82) - (-2279.7) kJ mol^{-1}
= +229.2 kJ mol^{-1}

$\Delta S^{\ominus}$ = 109 + 5 x 188.83 - 300.4 J K^{-1} mol^{-1}
= 752.2 J K^{-1} mol^{-1}

Therefore,

$T = \dfrac{299.2 \times 10^{3}\ \text{Jmol}^{-1}}{752.2\ \text{JK}^{-1}\text{mol}^{-1}}$ = **<u>397 K</u>**

3.43 DG = $SG_{prod\ -}$ SG_{reac}

= (-133 kJ - 394.36 kJ) - (-474 kJ) = **<u>-53.4 kJ</u>**

Additional problems

3.1 The enthalpy of vaporization, ΔH_{vap}, of carbon tetrachloride, CCl_4, at 298 K is 43 kJ mol^{-1}. If 1 mol CCl_4 (l) has an entropy of 214 J K^{-1} what is the entropy of 1 mol of its vapor at 298 K?

3.2 Calculate the change of entropy, DS, at 298 K for the reaction in which urea is formed from NH_3 and CO_2.

$$2NH_3(g) + CO_2(g) \rightarrow NH_2CONH_2(aq) + H_2O(l)$$

3.3 The reaction $CO_2(g) + H_2(g) \rightarrow CO(g) + H_2O(g)$ is nonspontaneous at 298 K but becomes spontaneous at higher temperatures. What can be concluded about DH and DS (assuming these parameters are not affected by temperature)?

3.4 K_a for acetic acid at 298 K is 1.754 x 10^{-5} and at 323 K, K_a = 1.633 x 10^{-5}. What are DH and DS for the reaction?

3.5 A sample of perfect gas occupies 10.0 L at 298 K and 1.00 atm. The gas is compressed isothermally to 1.0 L. What is the change in entropy?

3.6 Calculate DS for the reduction of aluminum oxide by hydrogen gas.

$$Al_2O_3(s) + 3H_2(g) \rightarrow 2Al(s) + 3H_2O(g)$$

3.7 For the reaction

$$4Fe(s) + 3O_2(g) \rightleftharpoons 2Fe_2O_3(s)$$

using data for $\Delta H_f^{\ominus}$ and $S^{\ominus}$ calculate the equilibrium constant for the reaction at 298 K.

3.8 Calculate the temperature for the conversion of monoclinic sulfur to rhombin sulfur.

	$\Delta H_f^\ominus$	$\Delta S^\ominus$
S(rhombic)	0 kJ	31.80 J K^{-1}
S(monoclinic)	0.33	32.6

3.9 A rought approximation is that for a chemical reaction

$\Delta S^\ominus \approx 100(Dn)$ J K^{-1} where Dn is the change in the amount (in moles) of gaseous components. use this approximation and the bond energies to estimate $\Delta G^\ominus$ for

$CO(g) + 2H_2(g) \rightarrow CH_3OH(g)$

3.10 A flask containing 1.00 mol of perfect gas at 4.00 bar and 298 K is connected to a flask containing 1.00 mol of perfect gas at 2.00 bar and 298 K. The gases are allowed to mix isothermally. What is the entropy change for the system?

3.11 As NaCl dissolves in water the solution cools slightly. $\Delta H^\ominus_{soln} = 3.88$ kJ. What is the sign of $\Delta S^\ominus_{soln}$? Calculate $\Delta S^\ominus_{soln}$ given $\Delta G_f^\ominus$(NaCl (*aq*)) = -393.1 kJ mol^{-1} and $\Delta G_f^\ominus$ (NaCl(*s*)) = -384.1 kJ mol^{-1}.

3.12 Calculate the change in entropy when a perfect gas is expanded to twice the volume and cooled to one half the initial temperature.

Chapter 4

Phase Equilibria

4.1 Use the perfect gas law to calculate the amount and the mass.

$$n = \frac{pV}{RT}, \quad n = \frac{m}{M}, \quad m = \frac{pVM}{RT}$$

(a) $$m = \frac{24\,\text{Torr} \times (101.8 \times 10^3\,\text{L}^3) \times (18.02\,\text{g mol}^{-1})}{(62.364\,\text{L Torr K}^{-1}\,\text{mol}^{-1}) \times (298.15\,\text{K})} = \underline{\mathbf{2.37\ kg}}$$

(b) $$m = \frac{(98\,\text{Torr}) \times (101.8 \times 10^3\,\text{L}^3) \times (78.11\,\text{g mol}^{-1})}{(62.364\,\text{L Torr K}^{-1}\,\text{mol}^{-1}) \times (298.15\,\text{K})} = \underline{\mathbf{41.9\ kg}}$$

(c) $$m = \frac{(1.7 \times 10^{-3}\,\text{Torr}) \times (101.8 \times 10^3\,\text{L}^3) \times (200.59\,\text{g mol}^{-1})}{(62.364\,\text{L Torr K}^{-1}\,\text{mol}^{-1}) \times (298.15\,\text{K})}$$
$$= \underline{\mathbf{1.87\ kg}}$$

4.2 The vapor pressure of ice at -5° C is 3.9 x 10^{-3} atm, or 3 Torr. The frost will sublime. A partial pressure of 3 Torr or more will ensure that the frost remains.

4.3 The volume decreases as the vapor is cooled from 400 K to 260 K. At the latter temperature the vapor condenses to a liquid and (if p is constant) there is a large decrease in volume. The liquid cools with only a small decrease in volume until the temperature reaches 273 K, when it freezes. The direction of the slope of the solid/liquid curve shows that the volume of the sample will then increase if the pressure is maintained. Ice remains at 260 K. There will be a pause in the rate of cooling at 373 K and a pause at 273 K.

4.4 Cooling from 400 K will cause the contraction of the gaseous sample until 273.16 K is reached, when the volume decreases by a large amount and solid ice is formed directly; liquid water may also form in equilibrium with the vapor and the solid.

4.5 (a) The gas expands. (b) The sample contracts but remains gaseous because 320 K is greater than the critical temperature. (c) The gas contracts and forms a liquid without the appearance of a discernable surface (d) The volume increases as the pressure on the liquid is reduced. (e) The liquid cools, then freezes, contracting as it does so. (f) The solid expands slightly as the pressure is reduced and sublimes when the pressure reaches about 5 atm. (g) The gas expands as it is heated at constant pressure.

4.6 Let A denote acetone and C chloroform. The mass of the sample is then

$$n_A M_A + n_C M_C = m$$

We also know that

$$x_A = \frac{n_A}{n_A + n_C} \quad \text{rearranging we get } (x_A - 1)n_A + x_A n_C = 0$$

Since $x_C = 1 - x_A$, we can substitute to obtain

$$-x_C n_A + x_A n_C = 0$$

On solving we obtain

$$n_A = \frac{x_A}{x_C} x n_C, \quad n_C = \frac{m x_C}{x_A M_A + x_C M_C}$$

Since $x_C = 0.4693, x_A = 1 - x_C = 0.5307$,

$$n_C = \frac{0.4693 \times 1000\,g}{(0.5307 \times 58.08 + 0.4693 \times 119.37)\,g\,mol^{-1}} = 5.404 \text{ mol}$$

$$n_A = \frac{0.5307}{0.4693} \times 5.404 = 6.111 \text{ mol}$$

The total volume, $V = n_A V_A + n_C V_C$ and

$V = 6.111 \text{ mol} \times 74.166 \text{ cm}^3 \text{ mol}^{-1} + 5.404 \times 80.235 \text{ cm}^3 \text{ mol}^{-1}$

$= \underline{\mathbf{886.8\ cm^3}}$

4.7 Check whether p_B / x_B is equal to a constant (K_B)

x	0.005	0.012	0.019
p/x	6×10^3	6.4×10^3	6.4×10^3 kPa

Hence, $K_B \sim \underline{\mathbf{6.4 \times 10^3\ kPa}}$

4.8 $p = p_A + p_B = x_A p_A^* + x_B p_B^* = x_A p_A^* + (1 - x_A) p_B^*$

Solving for x_A

$$x_A = \frac{p - p_B^*}{p_A^* - p_B^*}$$

For boiling under 0.50 atm (380 Torr) pressure, the combined vapor pressure must be 380 Torr, hence

$$x_A = \frac{380 - 150}{400 - 150} = \underline{\mathbf{0.920}},\ x_B = \underline{\mathbf{0.080}}$$

The composition of the vapor is given by

$$y_A = \frac{x_A p_A^*}{p_B^* + (p_A^* - p_B^*)x_A} = \frac{0.920 \times 400}{150 + (400 - 150) \times 0.920} = \underline{\mathbf{0.968}}$$

and $y_B = 1 - 0.968 = \underline{\mathbf{0.032}}$

4.9 Let B denote the benzene and A the solute, then

$p_B = x_B p_B^*$ and $x_B = \dfrac{n_B}{n_A + n_B}$

Hence $p_B = \dfrac{n_B p_B^*}{n_A + n_B}$

which solves to

$$n_A = \frac{n_B(p_B^* - p_B)}{p_B}$$

Then, since $n_A = m_A / M_A$, where m_A is the mass of A present,

$$M_A = \frac{m_A p_B}{n_B(p_B^* - p_B)} = \frac{m_A M_B p_B}{m_B(p_B^* - p_B)}$$

$$M_A = \frac{(0.125\,\text{g}) \times (78.11\,\text{g mol}^{-1}) \times (386\,\text{Torr})}{5.00\,\text{g} \times (400 - 386)\,\text{Torr}}$$

$= \underline{\mathbf{53.8\ g\ mol^{-1}}}$

4.10 $\Delta T = K_f m_B$ Where m_B is the molality of B and

$$m_B = \frac{28.0\,\text{g}}{M \times 750\,\text{g}} = \frac{0.0373}{M}$$

$$M = \frac{0.0373\ K_f}{\Delta T},$$

K_f = 30 K/(mol kg^{-1}) = 30 K kg mol^{-1}

$$M = \frac{(0.0373)\,x\,(30\,K\,kg\,mol^{-1})}{5.40\,K} = 0.207\ kg\ mol^{-1}$$

= **207 g mol^{-1}**

4.11 $K = \frac{[A_2]}{[A]^2}$, Initial A = n

fn = A_2 at equilibrium

A = (1 - 2f)n and the total amount of solute is (1-f)n. Therefore if the volume is V

$$K = \frac{fnV}{(1-2f)^2 n^2} = \frac{f}{(1-2f)^2 c} \quad \text{where } c = n/V$$

Vapor pressure, P is P = $x_{solute}P^*$

$$P = x_s P^* = \frac{n_s P^*}{n_A + n_{A_2} + n_s} = \frac{n_s P^*}{(1-f)n + n_s}$$

n_s = Vr with r = D/M

$$P = \frac{rP^*}{(1-f)c + r} \quad \text{rearranging } f = 1 - \frac{r(P^* - P)}{cP} \quad \text{and , finally}$$

$$\underline{K = \frac{1 - \frac{r(P^* - P)}{cP}}{c\left(1 - \frac{2r(P^* - P)}{cP}\right)^2}}$$

4.12 $\Pi V = n_B RT$ with $n_B / V \approx m_B p^{\ominus}$ for dilute solutions, with $p^{\ominus} = 10^3$ kg m^{-3}

$$\Delta T = K_f m_B \approx K_f \times \frac{\Pi}{RTp^{\ominus}}$$

Therefore with K_f = 1.86 K/(mol kg^{-1}) = 1.86 K kg mol^{-1}

$$\Delta T = \frac{(1.86\,\mathrm{K\,kg\,mol^{-1}}) \times (120 \times 10^3\,\mathrm{Pa})}{(8.314\,\mathrm{J\,K^{-1}\,mol^{-1}}) \times (300\,\mathrm{K}) \times (1.0 \times 10^3\,\mathrm{kg\,m^{-3}})} = 0.089\ \mathrm{K}$$

Therefore the solution will freeze at about **-0.09° C**

4.13 $p = xK$, $K = 1.25 \times 10^6$ Torr

$$x = \frac{n(CO_2)}{n(CO_2) + n(H_2O)} \approx \frac{n(CO_2)}{n(H_2O)}$$

Therefore

$$n(CO_2) \approx xn(H_2O) \text{ with } n(H_2O) = \frac{10^3\,\mathrm{g}}{18.02\,\mathrm{g\,mol^{-1}}} \text{ and } x = \frac{p}{K}$$

Substituting in

$$n(CO_2) \approx \frac{10^3\,\mathrm{g}}{18.02\,\mathrm{g\,mol^{-1}}} \times \frac{p}{1.26 \times 10^6\,\mathrm{Torr}}$$

$$\approx 4.4 \times 10^{-5}\ \mathrm{mol} \times (p/\mathrm{Torr})$$

(a) p = 4.0 kPa = 30.0 Torr,
hence $n(CO_2) = 4.4 \times 10^{-5}$ mol x 30.0 = 1.32×10^{-3} mol. The solution is therefore **1.3 mmol kg^{-1}** in CO_2.

(b) p = 100 kPa; since $n \propto p$, the solution is **33 mmol kg^{-1}** in CO_2.

4.14 $K(N_2) = 6.51 \times 10^7$ Torr and $K(O_2) = 3{,}30 \times 10^7$ Torr. Therefore as in problem 4.12, the amount of dissolved gas in l kg of water is

$$n(N_2) = \frac{10^3\,\text{g}}{18.02\,\text{g}\,\text{mol}^{-1}} \times \frac{p(N_2)}{6.51 \times 10^7\ \text{Torr}}$$

$= 8.52 \times 10^{-7}$ mol x (p/Torr)

For $p(N_2) = x\,p$ and p = 760 Torr
$n(N_2) = (8.52 \times 10^{-7}\ \text{mol}) \times (x) \times (760) = 6.48x \times 10^{-4}$ mol
and with $x = 0.78$,

$n(N_2) = 0.78 \times 6.48 \times 10^{-4}\ \text{mol} = 5.1 \times 10^{-4}$ mol
$= 0.51$ mmol

The molality of the solution is therefore approximately
5.1×10^{-4} mol kg^{-1}

For oxygen,

$$n(O_2) = \frac{10^3\,\text{g}}{18.02\,\text{g}\,\text{mol}^{-1}} \times \frac{p(O_2)}{3.30 \times 10^7\ \text{Torr}} = 1.68 \times 10^{-6}\ \text{mol} \times$$
(p/Torr)

For $p(O_2) = x\,p$ and p = 760 Torr

$n(O_2) = (1.68 \times 10^{-6}\ \text{mol}) \times (x) \times (760) = (1.28\ \text{mmol}) \times (x)$

when $x = 0.21$, $n(O_2) = 0.27$ mmol. Hence the solution will be **0.27 mmol Kg^{-1}** in O_2.

4.15 The amount of CO_2 in 1 kg of water is given by [Table 4.2]

$n(CO_2) = 4.4 \times 10^{-5}$ mol x (p/Torr)
substitute $p = 3.0 \times 760\ \text{Torr} = 2.28 \times 10^3$ Torr to give

$n(CO_2) = 4.4 \times 10^{-5}\ \text{mol} \times 2.28 \times 10^3 = 0.100$ mol

The molality of the solution is about **0.100 mol kg^{-1}** and the molar concentration about 0.17 M.

4.16 $\Delta T = K_f m_B = 1.86 \text{ K kg mol}^{-1} \times \dfrac{7.5\,\text{g}}{342.3\,\text{g mol}^{-1} \times 0.150\,\text{kg}}$

$= 0.27$ K

The freezing point will be approximately **-0.27 °C**.

4.17 $\Pi V = n_B RT$ so that $\Pi = \dfrac{mRT}{MV} = \dfrac{cRT}{M}$, $c = m/V$

$\Pi = \rho gh$ [hydrostatic pressure] so

$$h = \left(\frac{RT}{\rho gM}\right)c$$

A plot of h versus c will have a slope of $\dfrac{RT}{\rho gM}$. The slope of such a plot is 0.29 so

$$\frac{RT}{\rho gM} = \frac{0.29\,\text{cm}}{\text{g L}^{-1}} = 0.29 \times 10^{-2}\ \text{m}^4\ \text{kg}^{-1}$$

Therefore $M = \dfrac{RT}{\rho g \times 0.29 \times 10^{-2}\ \text{m}^4\ \text{kg}^{-1}}$

$$= \frac{(8.314\,\text{JK}^{-1}\text{mol}^{-1}) \times (298.15\,\text{K})}{(1.004 \times 10^3\,\text{kgm}^{-3}) \times (9.81\,\text{ms}^{-2}) \times (0.29 \times 10^{-2}\,\text{m}^4\,\text{kg}^{-1})}$$

$=$ **87 kg mol^{-1}**

4.18 As in problem 4.16, the data are plotted and the slope of the line is 1.78. Therefore

$$M = \frac{(8.3145\,\mathrm{JK^{-1}mol^{-1}}) \times (293.15\,\mathrm{K})}{(1.000 \times 10^3\,\mathrm{kg\,m^{-3}}) \times (9.81\,\mathrm{m\,s^{-2}}) \times (1.78 \times 10^{-2}\,\mathrm{m^4\,kg^{-1}})}$$

= **14 kg mol^{-1}**

4.19 The data are plotted in Fig. 4.1. From the graph, the vapor in equilibrium with a liquid of composition (a) $x_T = 0.25$ has y_T = **0.36** (b) for $x_O = 0.25$, $x_T = 0.75$ and y_T = **0.82**

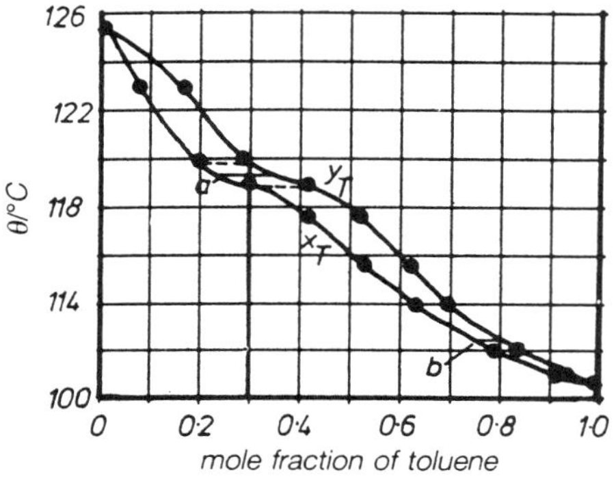

4.20

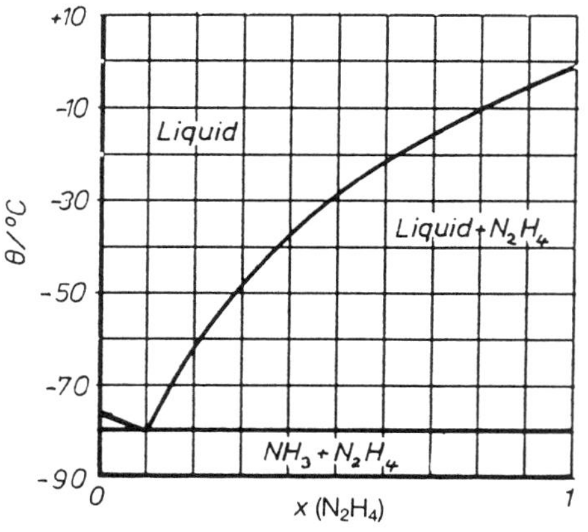

4.21 Refer to Fig. 4.31 of the text. At b_3 there are two phases with compositions $x_A = 0.18$ and $x_A = 0.70$; their abundances are in the ratio 0.13 [lever rule]. Since C = 2 and P = 2 we have F = 2 (such as p and x). On heating, the phases merge, and the single-phase region is encountered. Then F = 3 (such as p, T, and x). The liquid comes into equilibrium with its vapor when the isopeth cuts the phase line. At this temperature, and for all points up to b_1, C = 2 and P = 2, implying that F = 2. The whole sample is a vapor above b_1.

4.22 The phase diagrams are shown below

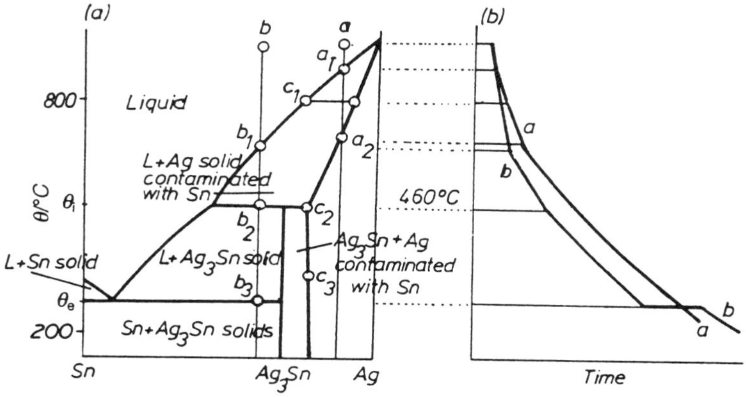

(a) Solid silver with dissolved tin begins to precipitate at a_1, and the sample solidifies completely at a_2. (b) Solid silver with dissolved tin begins to precipitate at b_1, and the liquid becomes richer in Sn. The peritectic reaction occurs at b_2, and as cooling continues Ag_3Sn is precipitated and the liquid becomes richer in tin. At b_3 the system has its eutectic composition (e) and freezes without further change.

4.23 The curves are shown in Fig. 4.2b. Note the eutectic halt for the isopleth b.

4.24 (a) The solubility of silver in tin at 800° C is determined by the point c_1 [at higher proportions of silver the system separates into two phases]. The point c_1 corresponds to **<u>80 percent</u>** silver by mass.

(b) See point c_2. The compound Ag_3Sn decomposes at this temperature.

(c) The solubility of Ag_3Sn in silver is given by point c_3 at 300° C.

4.25 The phase diagram is sketched below. (a) The mixture has a single liquid phase at all compositions. (b) When the composition reaches $x(C_6F_{10}) = 0.25$ the mixture separates into two liquid phases of composition x = 0.25 and 0.48. The relative amounts of the two phases change until the composition reaches x = 0.48. At all mole fractions greater than 0.48 in C_6F_{14} the mixture forms a single liquid phase.

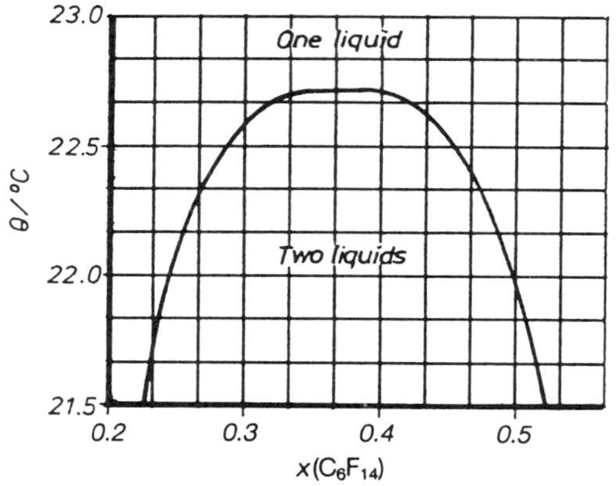

Additional problems

4.1 The entropy of vaporization of water is 40.7 kJ mol^{-1} and the vapor pressure of water at 298 K is 23.8 Torr. Calculate the vapor pressure of water at 325 K.

4.2 Calculate the expected vapor pressure at 25° C for a solution prepared by dissolving 200 g of sucrose (342.3 g mol^{-1}) in 700 mL of water.

4.3 Calculate the boiling point for the solution in Question 2 (K_b = 0.51).

4.4 A sample containing 20.0 g of an unknown substance is dissolved in 150.0 g of water. The resulting solution has a boiling point of 100.62° C. Calculate the molar mass of the unknown substance.

4.5 What mass of NaCl is required to lower the freezing point of 1.0 L of water to -10 °C?

4.6 The critical temperature of ammonia and nitrogen are 132° C and -147° C, respectively. Explain why ammonia can be liquified at room temperature by compression whereas liquefaction of nitrogen requires a lower temperature.

4.7 The vapor pressure of benzene is 100 Torr at 26.1° C and 400 Torr at 60.6° C. What is the boiling point of benzene at 760 Torr?

4.8 Describe the behavior of carbon dioxide gas when compressed at the following temperatures: (a) 20° C (b) -70° C and (c) 40° C. The triple point of CO_2 is -57° C and 5.1 atm and the critical point 31° C and 73 atm.

4.9 The osmotic pressure of blood at 37° C is 7.7 atm. An intraveneous solution must have the same osmotic pressure. What should the molar concentration of glucose solution be to give an osmotic pressure of 7.7 atm?

4.10 Why is a pressure cooker used to prepare foods at high altitude?

4.11 Some water is placed in a sealed gas container connected to a vacuum pump. The vacuum pump is turned on. Explain what change you would expect in the solution.

4.12 A soft drink is bottled so that at 25° C it contains CO_2 gas at a pressure of 5 atm over the liquid. Calculate the molar concentration of CO_2 in the liquid before and after opening the bottle. (Partial pressure of CO_2 in the atmosphere is 4.0 x 10^{-4} atm, Henry's Law constant for CO_2 is 32 L atm mol^{-1} at 25° C.)

4.13 An unknown protein has a molar mass of > 100,000 g mol^{-1}. Would it be more precise to determine the molar mass by freezing point depression or osmotic pressure?

4.14 After it is opened, wine is sometimes stored under nitrogen gas to prevent oxidation. The mole fraction of nitrogen in air is approximately 0.78. Calculate the molality of the nitrogen in wine open to the air and with nitrogen gas above it.

Chapter 5

Chemical equilibrium

5.1 a) $K = \dfrac{a_{COCl}a_{Cl}}{a_{CO}a_{Cl_2}}$

b) $K = \dfrac{a^2_{SO_3}}{a_{O_2}a^2_{SO_2}}$

c) $K = \dfrac{a^2_{HBr}}{a_{H_2}a^2_{Br}}$

d) $K = \dfrac{a^3_{O_2}}{a^2_{O_3}}$

5.2 $K = \dfrac{[C]}{[A][B]} = 0.224$

$$K = \frac{[A][B]}{[C]} = \frac{1}{0.224} = \underline{\mathbf{4.46}}$$

5.3 $K = \dfrac{[C]^2}{[A][B]} = 3.4 \times 10^4$

a) $K' = K^2 = (3.4 \times 10^4)^2 = \underline{\mathbf{1.16 \times 10^9}}$

b) $K'' = K^{1/2} = \underline{\mathbf{1.8 \times 10^2}}$

5.4 $I_2\,(g) \rightleftharpoons 2I(g)$

$$K = \frac{a_I^2}{a_{I_2}}$$

Initially, $[I_2] = \left(\frac{1.00\ g\ I_2}{1.0\ L}\right) x \left(\frac{1\ mol}{254\ g}\right) = 3.94 \times 10^{-3}$ mol L^{-1}

and at equilibrium,

$[I_2] = \left(\frac{0.83\ g\ I_2}{1.0\ L}\right) x \left(\frac{1\ mol}{254\ g}\right) = 3.27 \times 10^{-3}$ mol L^{-1}

and [I] = 2 x (mol I_2 reacted) = 2 x ($3.94 \times 10^{-3} - 3.27 \times 10^{-3}$)
= 1.34×10^{-3} mol L^{-1}

Therefore

$$K = \frac{[1.34 \times 10^{-3}]^2}{[3.3 \times 10^{-3}]} = \underline{\mathbf{5.4 \times 10^{-4}}}$$

5.5 $SbCl_5(g) \rightleftharpoons SbCl_3(g) + Cl_2(g)$

$$K = \frac{p_{SbCl_3} \cdot p_{Cl_2}}{p_{SbCl_5} \cdot p^{\ominus}} = 3.5 \times 10^{-4}$$

$$= \frac{(0.20) \cdot p_{Cl_2}}{(0.15)} = 3.5 \times 10^{-4}$$

$p_{Cl_2} = \underline{\mathbf{2.6 \times 10^{-4}}}$

5.6 $[PCl_5] = \frac{2.0\ g}{0.250\ L} x \frac{1\ mol}{206.22\ g} = 3.9 \times 10^{-2}$ mol L^{-1}

	$PCl_5(g) \rightleftharpoons$	$PCl_3(g)$ +	$Cl_2(g)$
(mol L^{-1})			
Initial	3.9×10^{-2}	0	0

Change	-x	+x	+x
Equil.	$(3.9 \times 10^{-2} - x)$	x	x

$$K = \frac{K}{(RT)^{\Delta n}} = \frac{0.36}{(0.0821) \times (400)^1} = 1.1 \times 10^{-2}$$

$$= \frac{x^2}{3.9 \times 10^{-2} - x}$$

Solving the quadratic

$x^2 + (1.1 \times 10^{-2})x - 1.5 \times 10^{-4} = 0$

$x = [PCl_3] = [Cl_2] = \underline{1.58 \times 10^{-2}}\ \underline{\text{mol L}^{-1}}$

$[PCl_5] = 3.9 \times 10^{-2} - x = \underline{\mathbf{2.3 \times 10^{-2}}}\ \underline{\mathbf{mol\ L^{-1}}}$

b) % decomposition =

$$\frac{\text{amount decomposed}}{\text{initial amount}} \times 100\% = \frac{1.6 \times 10^{-2}\ \text{mol L}^{-1}}{3.9 \times 10^{-2}\ \text{mol L}^{-1}} \times 100\% = \underline{\mathbf{42\ \%}}$$

5.7	$N_2(g)$ +	$3H_2(g)$ $\rightleftharpoons$	$2NH_3(g)$
Initial /atm	0.020	0.020	0
Change	-x	-3x	+2x
Equil	0.020 - x	0.020-3x	2x

$$K = \frac{p_{NH_3}^2 \cdot p^{\ominus}}{p_{N_2} \cdot p_{H_2}^3} = 0.036$$

$$0.036 = \frac{4x^2}{(0.02)(0.02)^3}$$

$x = 3.8 \times 10^{-5}$

Therefore $2x = p_{NH_3}$ = **7.6×10^{-5} bar**
and $p_{N_2} \approx$ **0.020 bar** and $p_{H_2} \approx$ **0.020 bar**

5.8 $K = \dfrac{(p_{NO_2})^2}{(p_{N_2O_4}) \cdot p^{\ominus}}$

$$N_2O_4(g) \rightleftharpoons 2NO_2(g)$$

At equilibrium 1- a 2a

$p_{NO_2} = x_{NO_2} p = 2ap$ and similarily, $p_{N_2O_4} = (1 - a)p$

$$K_p = \frac{(2\alpha p)^2}{(1-\alpha)p} = \frac{4\alpha^2 p}{(1-\alpha)}$$

When a << 1 then $K \approx 4\alpha^2 p / p^{\ominus}$ and $\alpha \propto \dfrac{1}{\sqrt{(p/p^{\ominus})}}$

5.9 a)

conjugate

H_2SO_4 + H_2O → H_3O^+ + HSO_4^-
acid$_1$ base$_2$ acid$_2$ base$_1$

conjugate

b)

conjugate

HF + H_2O → H_3O^+ + F^-
acid$_1$ base$_2$ acid$_2$ base$_1$

conjugate

c)

conjugate

$C_6N_5NH_3^+$ + H_2O → H_3O^+ + $C_6H_5NH_2$

acid$_1$ base$_2$ acid$_2$ base$_1$
conjugate

d) conjugate
$H_2PO_4^-$ + H_2O → H_3O^+ + HPO_4^{2-}
acid$_1$ base$_2$ acid$_2$ base$_1$
conjugate

e) conjugate

HCOOH + H_2O → H_3O^+ + HCO_2^-
acid$_1$ base$_2$ acid$_2$ base$_1$
conjugate

f) conjugate
$NH_2NH_3^+$ + H_2O → H_3O^+ + NH_2NH_2
acid$_1$ base$_2$ acid$_2$ base$_1$
conjugate

5.10 a) $CH_3CH(OH)COOH + H_2O \rightarrow$
$CH_3CH(OH)COO^- + H_3O^+$

b) $HOOCH_2C(NH_2)COOH + H_2O \rightarrow$
$HOOCH_2C(NH_2)COO^- + H_3O^+$

$HOOCH_2C(NH_2)COO^- + H_2O \rightarrow$
$^-OOCH_2C(NH_2)COO^- + H_3O^+$

c) $NH_2CH_2COOH + H_2O \rightarrow$
$^+NH_3CH_2COO^- + H_2O$

d) $HOOCCOOH + H_2O \rightarrow$
$HOOCCOO^- + H_3O^+$

$HOOCCOO^- + H_2O \rightarrow {}^-OOCCOO^- + H_3O^+$

5.11 a) $K_w = 2.5 \times 10^{-14} = [H_3O^+][OH^-] = x^2$

$[H_3O^+] = \sqrt{2.5 \times 10^{-14}} = 1.6 \times 10^{-7}$ mol L^{-1}

pH = -log$[H_3O^+]$ = **<u>6.8</u>**

b) $[OH^-] = [H_3O^+] = 1.6 \times 10^{-7}$ mol L^{-1}

pOH = -log$[OH^-]$ = **<u>6.8</u>**

5.12 $K_{w,D} = [D_3O^+][OD^-] = 1.35 \times 10^{-15}$

a) $D_2O + D_2O \rightleftharpoons D_3O^+ + OD^-$

b) $K_{w,D} = [D_3O^+][OD^-] = 1.35 \times 10^{-15}$
$pK_w = -\log K_w$ = **<u>14.87</u>**

c) $[D_3O^+] = [OD^-] = \sqrt{1.35 \times 10^{-15}}$ = **<u>3.67×10^{-8} mol L^{-1}</u>**

d) pD = - log(3.67×10^{-8}) = **<u>7.43</u>** = pOD

e) pD + pOD = $pK_w(D_2O)$ = **<u>14.87</u>**

5.13 a) pH = -log (1.5×10^{-5}) = **<u>4.82</u>**

pOH = 14.00 - pH = 14.0 - 4.82 = **<u>9.18</u>**

b) pH = -log(1.5×10^{-3}) = **<u>2.82</u>**

pOH = 14.00 - 2.82 = **<u>11.18</u>**

c) pH = -log (5.1×10^{-14}) = 13.30

but $[H_3O^+] = 1.0 \times 10^{-7}$ from water, **<u>pH = pOH = 7.0</u>**

d) pH = -log(5.01×10^{-5}) = **<u>4.30</u>**

pOH = 14.00 - 4.30 = **<u>9.70</u>**

5.14 a) amount (moles) H_3O^+ = (0.025 L) x (0.144 mol L^{-1})
= 3.60 x 10^{-3} mol

amount (moles) OH^- = (0.025 L) x (0.125 mol L^{-1})
= 3.125 x 10^{-3} mol

excess H_3O^+ = (3.6 x 10^{-3} - 3.125 x 10^{-3}) = 0.475 x 10^{-3} mol H_3O^+

$$[H_3O^+] = \frac{4.75 \times 10^{-4}\ \text{mol}}{0.050\ \text{L}} = \mathbf{9.50 \times 10^{-3}\ mol\ L^{-1}}$$

pH = - log(0.050) = **2.02**

b) amount of H_3O^+ = (0.025 L) x (0.15 mol L^{-1})
= 3.75 x 10^{-3} mol H_3O^+

amount of OH^- = (0.035 L) x (0.15 mol L^{-1})
= 5.25 x 10^{-3} mol OH^-

excess OH^- = (5.25 x 10^{-3} - 3.75 x 10^{-3}) = 1.50 x 10^{-3} mol OH^-

$$[OH^-] = \frac{1.50 \times 10^{-3}\ \text{mol}}{0.060\ \text{L}} = \mathbf{0.025\ mol\ L^{-1}}$$

pOH = -log(0.025) = 1.60

pH = 14.00 - 1.30 = **12.40**

c) Amount of H_3O^+ = (0.0212 L) x (0.22 mol L^{-1}) = 4.66 x 10^{-3} mol H_3O^+
amount of OH^- = (0.010 L) x (0.30 mol L^{-1})
= 3.0 x 10^{-3} mol OH^-

$$\text{Concentration of excess } H^+ = \frac{1.66 \times 10^{-3}\ \text{mol}}{0.031\ \text{L}} = 5.35 \times 10^{-2}\ \text{M}$$

pH = **1.27**

5.15

a) acidic; $NH_4^+(aq) + H_2O(l) \rightleftharpoons H_3O^+(aq) + NH_3(aq)$

b) basic; $H_2O(l) + CO_3^{2-}(aq) \rightleftharpoons HCO_3^-(aq) + OH^-(aq)$

c) basic; $H_2O(l) + F^-(aq) \rightleftharpoons HF(aq) + OH^-(aq)$

d) neutral

e) acidic; $[Al(H_2O)_6]^{3+}(aq) + H_2O(l) \rightleftharpoons [Al(H_2O)_5OH]^{2+}(aq) + H_3O^+(aq)$

f) acidic; $[Co(H_2O)_6]^{2+}(aq) + H_2O(l) \rightleftharpoons [Co(H_2O)_5OH]^{+}(aq) + H_3O^+(aq)$

5.16 a) Calculate the molar concentration of $KC_2H_3O_2$

$$(8.4\text{ g}) \times \left(\frac{1\text{ mol}}{98.146\text{ g}}\right) \times \left(\frac{1}{0.250\text{ L}}\right) = 0.342\text{ mol L}^{-1}$$

	$H_2O(l)$	+ $C_2H_3O_2^-(aq)$	$\rightleftharpoons$ $HC_2H_3O_2(aq)$	+ OH^-
Initial	--	0.342	0	0
Change	--	-x	+x	+x
Equil.	--	0.342 - x	x	x

$$K_b = \frac{K_w}{K_a} = \frac{1.0 \times 10^{-14}}{1.8 \times 10^{-5}} \approx \frac{x^2}{0.342}$$

$[OH^-] = 1.4 \times 10^{-5}\text{ mol L}^{-1}$

pOH = 4.86

pH = 14.00 - 4.86 = **9.14**

b) $(3.55\text{ g}) \times \left(\frac{1\text{ mol}}{97.9\text{ g}}\right) \times \left(\frac{1}{0.100\text{ L}}\right) = 0.383\text{ mol L}^{-1}\ NH_4Br$

	$NH_4^+(aq)$ +	$H_2O(l)$ ⇌	$NH_3(aq)$ +	$H_3O^+(aq)$
Initial	0.383	----	0	0
Change	-x	----	+x	+x
Equil	0.383 - x	----	x	x

$$\frac{1.0 \times 10^{-14}}{1.8 \times 10^{-5}} \approx \frac{x^2}{0.383}$$

$[H_3O^+] = 1.46 \times 10^{-5}\text{ mol L}^{-1}$

pH = $-\log(1.46 \times 10^{-5})$ = **4.83**

c) HBr is a strong acid and is therefore essentially completely ionized in aqueous solution. Therefore none of the Br^- is protonated.

5.17 a) K_a = **8.3×10^{-4}**
$pK_a = 3.08$

b)

	$HL(aq)$ +	$H_2O(l)$ ⇌	$H_3O^+(aq)$ +	$L^-(aq)$
Initial	x	----	----	2x
Change	-y	----	+y	+y

$$K_a = \frac{[H_3O^+][L^-]}{[HL]} = \frac{[y][y+x]}{[2x-y]} \approx \frac{[y][x]}{[2x]} = 8.4 \times 10^{-4}$$

y = 2(8.4 x 10^{-4} = 1.68 x 10^{-3} mol L^{-1} = [H_3O^+]

pH = **2.8**

5.18

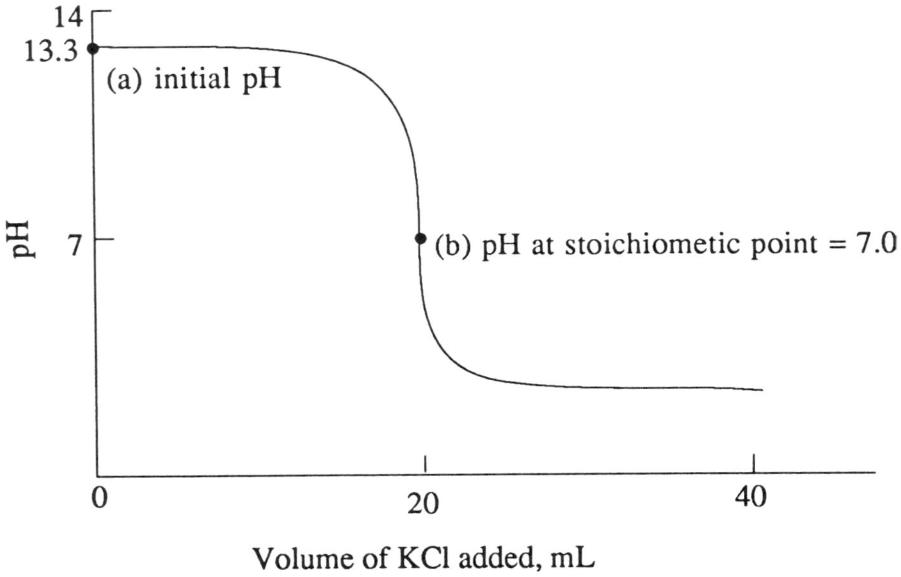

Initial pH = 14.00 - (-log 0.30) = **13.48**

At the stoichiometric point for a strong acid and strong base the pH = 7.0

Volume of HCl at the stoichiometric point

0.025 L x 0.15 mol L^{-1} x 2 mol OH^-
= mol OH^- = mol H^+ = 7.5 x 10^{-3}

7,5 x 10^{-3} mol/0.22 mol L^{-1} = 0.34 L volume at stoichiometric point

5.19 a) $C_6H_5COOH + H_2O \rightleftharpoons H_3O^+ + C_6H_5COO^-$

	C_6H_5COOH	H_2O	H_3O^+	$C_6H_5COO^-$
Init. (mol L^{-1})	0.250	---	0	0
Change	-x	---	+x	+x
Equil.	0.250- x	---	x	x

$$K_a \approx \frac{x^2}{0.250} = 6.5 \times 10^{-5}$$

$x = 4.03 \times 10^{-3}$

$$\text{percentage ionized} = \frac{4.0 \times 10^{-3}}{0.250} \times 100\% = \underline{\mathbf{1.6\ \%}}$$

b) $H_2O + NH_2NH_2 \rightleftharpoons NH_2NH_3^+ + OH^-$

	H_2O	NH_2NH_2	$NH_2NH_3^+$	OH^-
Init.	---	0.150	0	0
Change	---	-x	+x	+x
Equil.	---	0.150 - x	x	x

$$K_b = \frac{[NH_2NH_3^+][OH^-]}{[NH_2NH_2]} \approx \frac{x^2}{0.150} = 1.7 \times 10^{-6}$$

$x = 5.0 \times 10^{-4}$

$$\text{percentage ionized} = \frac{5.0 \times 10^{-4}}{0.150} \times 100\ \% = \underline{\mathbf{0.33\ \%}}$$

c) $(CH_3)_3N \quad H_2O \rightleftharpoons (CH_3)_3NH^+ + OH^-$

	$(CH_3)_3N$	H_2O	$(CH_3)_3NH^+$	OH^-
Init.	0.112	---	0	0
Change	-x	---	+x	+x

Equil. 0.112 - x --- x x

$$K_b = \frac{[(CH_3)_3NH][OH^-]}{[(CH_3)_3N]} \approx \frac{x^2}{0.112} = 6.5 \times 10^{-5}$$

$x = 2.7 \times 10^{-3}$

$$\text{percentage ionized} = \frac{2.7 \times 10^{-3}}{0.112} \times 100\% = \underline{\mathbf{2.4\ \%}}$$

5.20 See the above for detailed information on setting up the equilibrium calculations for weak acids.

$$K_a = 8.4 \times 10^{-4} = \frac{[H_3O^+][CH_3CH(OH)CO_2^-]}{[CH_3CH(OH)COOH]} = \frac{x^2}{0.120 - x} \approx \frac{x^2}{0.120}$$

Assuming x is small relative to 0.120.
$x = [H_3O^+] = 0.010$

pH = -log(0.010) = 2.00

pOH = 14.00 - 2.00 = 12.00

$$\text{Fraction ionized} = \frac{0.010}{0.120} = \underline{\mathbf{0.083}}$$

Without the approximation the quadratic equation must be solved

b)

$$8.4 \times 10^{-4} = \frac{x^2}{1.4 \times 10^{-4} - x}$$

$x^2 + 8.4 \times 10^{-4}x - 1.18 \times 10^{-8} = 0$

$x = 1.4 \times 10^{-5}$ [negative root is not possible]

pH = -log(1.4 x 10^{-5}) = 4.85

pOH = 14.00 - 4.85 = 9.15

$$\text{Fraction ionized} = \frac{1.4 \times 10^{-5}}{1.4 \times 10^{-4}} = \mathbf{\underline{0.10}}$$

$$\text{c) } K_a = \frac{[H_3O^+][C_6H_5SO_3^-]}{[C_6H_5SO_3H]} = \frac{x^2}{(0.10) - x} = 0.20$$

$x^2 + 0.20x - 0.02 = 0$

$$x = \frac{-(0.2) \pm \sqrt{(0.2)^2 - 4\times(-0.02)}}{2} = 0.073$$

pH = -log(0.073) = **<u>1.1</u>**

pOH = 14.0 - 1.1 = **<u>12.9</u>**

$$\text{Fraction ionized} = \frac{0.073}{0.10} = \mathbf{\underline{0.73}}$$

5.21 a) K_a = 7.2 x 10^{-10}

$$= \frac{[H_3O^+][B(OH)_4^-]}{[B(OH)_3]} = \frac{x^2}{(1.0\times10^{-4}) - x} \approx \frac{x^2}{1.0\times10^{-4}}$$

x = [H_3O^+] = 2.7 x 10^{-7} mol L^{-1}

pH = -log(2.7 x 10^{-7}) = 6.57

This value of [H_3O^+] is not much different from water and therefore we should consider the contribution from the ionization of water. We should therefore simultaneously consider both equilibria.

$$B(OH)_3 \quad + \quad 2H_2O \rightleftharpoons \quad H_3O^+ \quad + \quad B(OH)_4^-$$

Equil	$1.0 \times 10^{-4} - x$	----	x	y
	$2H_2O$	$\rightleftharpoons$	H_3O^+ +	OH^-
Equil.	---		x	z

Since there are now two contributions to $[H_3O^+]$, it is no longer equal to $[B(OH)_4^-]$, nor is it equal to $[OH^-]$ as in pure water. To avoid a cubic equation, x will be ignored relative to 1.0×10^{-4} mol L^{-1}. Let a = initial concentration of $B(OH)_3$

$$K_a = 7.2 \times 10^{-10} = \frac{xy}{a} \quad \text{or } y = aK_a/x$$

$$K_w = 1.0 \times 10^{-14} = xz$$

Electroneutrallity requires x = y + z or z = x - y

$$K_w = xz = x(x - y)$$

Substituting in for y from above

$$K_w = x\left(x - \frac{aK_a}{x}\right)$$

$$x^2 - aK_a = K_w$$

$$x = \sqrt{K_w + aK_a} = \sqrt{(1.0 x 10^{-14}) + (1.0 x 10^{-4}) x (7.2 x 10^{-10})}$$

$$x = 2.86 \times 10^{-7} \text{ mol L}^{-1}$$

pH = -log(2.86 x 10^{-7}) = **6.5**

This value is slightly, but measurably different from the value 6.57 obtained by ignoring the contribution from water.

b) In this case the second ionization can be ignored; $K_{a2} << K_{a1}$

$$K_{a1} = 7.6 \times 10^{-3} = \frac{x^2}{(0.015) - x}$$

$x^2 + 7.6 \times 10^{-3}x - 1.14 \times 10^{-4} = 0$

$$x = [H_3O^+] = \frac{-(7.6 \times 10^{-3}) \pm \sqrt{(7.6 \times 10^{-3})^2 - (-4.56 \times 10^{-4})}}{2}$$

$= 7.53 \times 10^{-3}$ mol L^{-1}

pH = -log(7.53×10^{-3}) = **2.1**

c) The second ionization can be ignored; $K_{a2} << K_{a1}$

$$K_{a1} = 1.5 \times 10^{-2} = \frac{x^2}{(0.10) - x}$$

$x^2 + 1.5 \times 10^{-2}x - 1.5 \times 10^{-3} = 0$

$$x = [H_3O^+] = \frac{-(1.5 \times 10^{-2}) \pm \sqrt{(1.5 \times 10^{-2})^2 - (-6.0 \times 10^{-3})}}{2}$$

= 0.032 mol L^{-1}

pH = -log(0.032) = **1.5**

5.22 Would act as a buffer at pH $\approx pK_a$ or pH near 8.3.

5.23 a) $pH = pK_a - \log \frac{[Acid]}{[Base]}$

$7.00 = 2.20 - \log \frac{[Acid]}{[Base]}$

$\log \frac{[Acid]}{[Base]} = -4.8$

$\frac{[Acid]}{[Base]}$ = **1.58×10^{-5}**

b) $\log \frac{[\text{Acid}]}{[\text{Base}]} = 0$

$\frac{[\text{Acid}]}{[\text{Base}]} = \underline{\mathbf{1}}$

c) $\log \frac{[\text{Acid}]}{[\text{Base}]} = 0.7$

$\frac{[\text{Acid}]}{[\text{Base}]} = \underline{\mathbf{5.0}}$

5.24 a)

	$H_2C_2O_4$	+ H_2O $\rightleftharpoons$	H_3O^+ +	$HC_2O_4^-$
Initial	0.15	---	0	0
Change	-x	---	+x	+x
Equil.	0.15 - x	---	x	x

$K_{a1} = 5.9 \times 10^{-2}$, $K_{a2} = 6.5 \times 10^{-5}$

The second ionization can be ignored in the calculation of $[H_3O^+]$. but not in the calculation of $C_2O_4^-$.

$K_{a1} = 5.9 \times 10^{-2} = \frac{x^2}{0.15 - x}$

$x = \frac{-(0.059) \pm \sqrt{(0.059)^2 - 4(0.00885)}}{2} = 0.060 \text{ mol L}^{-1} = [H_3O+]$

$[OH^-] = 1.7 \times 10^{-13}$ mol L^{-1}

$[H_2C_2O_4] = 0.15 - 0.060 = 0.090$ mol L^{-1}

$HC_2O_4^- + H_2O \rightleftharpoons H_3O^+ + C_2O_4^-$

Initial	0.060	---	0.060	0
Change	-x	---	+x	+x
Equil.	0.060 - x		0.060 + x	x

$K_{a2} = 6.5 \times 10^{-5} = \frac{(0.06+x)x}{(0.60-x)} \approx x$ [because x is small]

$x = [C_2O_4^{2-}] = 6.5 \times 10^{-5}$ mol L^{-1} and

$[HC_2O_4^-] = 0.06 - x = 0.060 - 0.000065 =$ **0.060 mol L^{-1}**

b) Similar to part a, K_{a2} can be ignored in the first part of the calculation and

$K_{a1} = 1.3 \times 10^{-7} = \frac{x^2}{0.065 - x} \approx \frac{x^2}{0.065}$,

$[H_2S] =$ **0.065 mol L^{-1}**

$x = [H_3O^+] = [HS^-] =$ **9.2×10^{-5} mol L^{-1}**

$[OH^-] = \frac{1.0 \times 10^{-14}}{9.2 \times 10^{-5}} =$ **1.1×10^{-10} mol L^{-1}**

For the second ionization,
$K_{a2} = 7.1 \times 10^{-15} = x = [S^{2-}]$

$[S^{2-}] =$ **7.1×10^{-15} mol L^{-1}**

5.25 a) 0.10 M CH_3COOH

$K_a = 1.8 \times 10^{-5} = \frac{[H_3O^+][CH_3CO_2^-]}{[CH_3COOH]} = \frac{x^2}{(0.10) - x} \approx \frac{x^2}{0.10}$

$x^2 = 1.8 \times 10^{-6}$

$x = 1.3 \times 10^{-3}$ mol L^{-1}

initial pH = -log (1.3×10^{-3}) = **<u>2.87</u>**

b) n (CH_3COOH) = (0.0250 L) x (0.10 mol L^{-1})
= 2.50×10^{-3} mol

n (NaOH) = (0.0100 L) x (0.10 mol L^{-1}) - 1.0×10^{-3} mol

Therefore $[CH_3COOH] = \dfrac{1.5 \times 10^{-3}}{0.0350} = 4.29 \times 10^{-2}$ mol L^{-1}

and $[CH_3COO^-] = \dfrac{1.0 \times 10^{-3}}{0.0350} = 2.86 \times 10^{-2}$ mol L^{-1}

Consider the equilibrium

	CH_3COOH	+ H_2O ⇌	H_3O^+	+ $CH_3CO_2^-$
Init.	4.29×10^{-2}	----	0	2.86×10^{-2}
Change	-x	----	+x	+x
Equil.	4.29×10^{-2} - x	----	x	2.86×10^{-2} + x

$1.8 \times 10^{-5} = \dfrac{(2.86 \times 10^{-2} + x)(x)}{(4.29 \times 10^{-2} - x)}$; assume +x and -x are negligible

$[H_3O^+] = x = 2.7 \times 10^{-5}$ mol L^{-1}

pH = -log(2.7×10^{-5}) = **<u>4.56</u>**

c) Since acid and base concentrations are equal, their volumes are equal at the stoichiometric point. Therefore 25.0 mL NaOH to reach the stoichiometric point and <u>12.5 mL</u> to reach the half stoichiometric point.

d) At halfway to the stoichiometric point, pH = pK_a and

pH = -log(1.8 x 10^{-5}) = **4.74**

e) **25.0 mL** [see part c]

f) The final pH is that of 0.050 M $NaCH_3CO_2$

$$K_b = \frac{K_w}{K_a} = \frac{1.00 \times 10^{-14}}{1.8 \times 10^{-5}} \frac{x^2}{(0.050) - x} \approx \frac{x^2}{0.050}$$

$x^2 = 2.8 \times 10^{-11}$

x = 5.5 x 10^{-6} mol L^{-1} = [OH^-]

pOH = 5.28, pH = 14.00 - 5.28 = **8.72**

5.26 $K_a = \frac{[H_3O^+][CH_3CO_2^-]}{[CH_3COOH]}$

$$pH = pK_a + \log \frac{[CH_3CO_2^-]}{[CH_3COOH]}$$

a) $pH = pK_a + \log \frac{[0.10]}{[0.10]} = pK_a$ = -log(1.8 x 10^{-5}) = **4.74**

b) 3.3 mmol NaOH = 3.3 x 10^{-3} mol OH^- [strong base]
produces 3.3 x 10^{-3} mol CH_3COO^- from CH_3COOH

Initially n (CH_3COOH) = n (CH_3COO^-) = 0.10 mol L^{-1} x 0.100 L = 1.0 x 10^{-2} mol

After adding NaOH

$$[CH_3COOH] = \frac{1.0 \times 10^{-2} - 3.3 \times 10^{-3}}{0.10\ L} = 6.7 \times 10^{-2}\ mol\ L^{-1}$$

$$[CH_3COO^-] = \frac{1.0 \times 10^{-2} + 3.3 \times 10^{-3}}{0.10\ L} = 0.13\ mol\ L^{-1}$$

$$pH = 4.74 + \log \frac{0.13}{0.067} = \underline{\mathbf{5.04}},$$

Change in pH = **<u>0.30</u>**

c) 6.0 mmol HNO_3 = 6.0 x 10^{-3} mmol H_3O^+ [strong acid]

produces 6.0 x 10^{-3} mol CH_3COOH from CH_3COO^-

after adding HNO_3 [see (b) above]

$$[CH_3COOH] = \frac{(1.0 \times 10^{-2}) + (6.0 \times 10^{-3})\ \text{mol}}{0.100\ \text{L}} = 0.16\ \text{mol L}^{-1}$$

$$[CH_3COO^-] = \frac{(1.0 \times 10^{-2}) - (6.0 \times 10^{-3})\ \text{mol}}{0.100\ \text{L}} = 4.0 \times 10^{-2}\ \text{mol L}^{-1}$$

$$pH = 4.74 + \log \frac{4.0 \times 10^{-2}}{0.16} = 4.74 - 0.60 = \underline{\mathbf{4.14}},$$

change in pH = **<u>-0.60</u>**

5.27 These will be effective buffers at pH's near the pK_a.

a) **<u>pH 2 - 4</u>**

b) **<u>pH 3 - 5</u>**

c) **<u>pH 11 - 13</u>**

d) **<u>pH 6 - 8</u>**

e) **<u>pH 5 - 7</u>**

5.28 At the halfway point, pH = pK_a = **<u>4.66</u>**
and K_a = **<u>2.19 x 10^{-5}</u>**

$$K_a = 2.19 \times 10^{-5} = \frac{x^2}{(0.015 - x)} \approx \frac{x^2}{(0.015)}$$

$x = 5.7 \times 10^{-4}$

pH = **3.24**

5.29 $NH_4^+ + H_2O \rightleftharpoons NH_3 + H_3O^+$

$$K_a = \frac{[NH_3][H_3O^+]}{[NH_4^+]} = 5.6 \times 10^{-10}$$

$$= \frac{x^2}{0.15 - x} \approx \frac{x^2}{0.15}$$

$x = 9.2 \times 10^{-6}$

pH = $-\log(9.2 \times 10^{-6})$ = **5.04**

b) $CH_3COO^- + H_2O \rightleftharpoons CH_3COOH + OH^-$

$$K_b = 5.6 \times 10^{-10} = \frac{x^2}{0.15 - x} \approx \frac{x^2}{0.15}$$

$x = 9.2 \times 10^{-5} = [OH^-]$

pOH = $-\log(9.2 \times 10^{-5})$ = 5.04

pH = 14 - 5.04 = **8.96**

c) Use the reverse of the above reaction.

$$K_a = 1.8 \times 10^{-5} = \frac{x^2}{0.15 - x} \approx \frac{x^2}{0.15}$$

$x = 1.64 \times 10^{-3} = [H_3O^+]$

pH = -log(1.64 x 10^{-3}) = **2.78**

5.30 At the stoichiometric point the solution will consist of the weak base and Na^+ ions. To calculate the pH, first calculate the dilution by adding base.

mol $CH_3CH(OH)COO^-$ = 0.025 L x 0.10 mol L^{-1}

= 2.5 x 10^{-3} mol

volume of base required = (2.5 x 10^{-3} mol)/(0.175 mol L^{-1})

= 0.014 L

Total volume is 0.025 + 0.014 L = 0.039 L

[$CH_3CH(OH)COO^-$] = (2.5 x 10^{-3} mol)/(0.039 L) = 0.064 M

$$K_b = 1.2 \times 10^{-11} = \frac{x^2}{0.064 - x} \approx \frac{x^2}{0.064}$$

x = 8.8 x 10^{-7}

pOH = -log(8.8 x 10^{-7}) = 6.06

pH = 14 - 6.06 = **7.94**

5.31 The initial pH is calculated in problem 5.26 (b). Use the Henderson-Hasselbach equation to calculate other pH's. The inflection point will be at pH = pK = 4.74.

5.32 a) H_3PO_4 and NaH_2PO_4

b) NaH_2PO_4 and Na_2HPO_4

5.33 a) $K_s = [Ag^+][I^-]$

b) $K_s = [Hg_2^{+2}][S^{2-}]$

c) $K_s = [Fe^{3+}][OH^-]^3$

d) $K_s = [Ag^+]^2[CrO_4^{2-}]$

5.34 $BaSO_4(s) \rightleftharpoons Ba^{2+}(aq) + SO_4^{2-}(aq)$

Initial	0	0
Change	+x	+x
Equil.	x	x

$K_s = [Ba^{2+}][SO_4^{2-}]$

$1.1 \times 10^{-10} = x^2$

x = **1.0 x 10^-5 mol L^-1** → $x = \mathbf{1.0 \times 10^{-5}\ mol\ L^{-1}}$

b) $K_s = [Ag^+]^2[CO_3^{2-}]$

Since 2 mol of Ag^+ formed for 1 mol CO_3^{2-}

$K_s = (2x)^2(x) = 6.2 \times 10^{-12} = 4x^3$

$x = \mathbf{1.1 \times 10^{-4}\ mol\ L^{-1}}$

c) $K_s = [Fe^{3+}][OH^-]^3$

$2.0 \times 10^{-39} = (x)(3x)^3 = 27x^4$

$x = \mathbf{9.28 \times 10^{-11}\ mol\ L^{-1}}$

d) $K_s = [Hg_2^{2+}][Cl^-]^2 = 1.3 \times 10^{-18} = 4x^3$

$x = \mathbf{6.8 \times 10^{-7}\ mol\ L^{-1}}$

5.35 a) $AgBr(s) \rightleftharpoons Ag^+(aq) + Br^-(aq)$

Initial	--	0	0.0014
Change	--	+x	+x
Equil.	--	x	0.0014 + x

$K_s = [Ag^+][Br^-] = 7.7 \times 10^{-13} = (x) \times (0.0014 + x) \approx (x) \times (.0014)$

x = **5.5×10^{-10} mol L^{-1}** = $[Ag^+]$ = molar solubility of AgBr in 0.0014 M NaBr.

b) $MgCO_3(s) \rightleftharpoons Mg^{2+}(aq) + CO_3^{2-}(aq)$

Initial	--	0	1.1×10^{-5}
Change	--	+x	+x
Equil.		x	$1.1 \times 10^{-5} + x$

$K_s = [Mg^{2+}][CO_3^{2-}] = (x) \times (1.1 \times 10^{-5} + x) = 1.0 \times 10^{-5}$

Assume 1.1×10^{-5} negligible

$x^2 = 1.0 \times 10^{-5}$

x = **3.2×10^{-3} mol L^{-1}**

c) $PbSO_4(s) \rightleftharpoons Pb^{2+}(aq) + SO_4^{2-}(aq)$

Init.	0	0	0.10
Change		+x	+x
Equil.		x	0.10 + x

$K_s = [Pb^{2+}][SO_4^{2-}] = 1.6 \times 10^{-8} = (x) \times (0.10 + x) = 0.10\,(x)$

x = **1.6×10^{-7} mol L^{-1}**

d) $Ni(OH)_2(s) \rightleftharpoons Ni^{2+}(aq) + 2OH^-(aq)$

Initial	2.7×10^{-5}	0
Change	+x	+2x
Equil.	$2.7 \times 10^{-5} + x$	2x

$K_s = [Ni^{2+}][OH^-]^2 = 6.5 \times 10^{-18} = (x + 2.7 \times 10^{-5}) \times (2x)^2$

assume x in $(x + 2.7 \times 10^{-5})$ is negligible

$6.5 \times 10^{-18} = (7.4 \times 10^{-5})4x^2$
x = **2.45×10^{-7} mol L^{-1}**

5.36 $G' + NH_4^+ \rightarrow G$ DG = +15.7 kJ mol^{-1}

ATP -> ADP + P_1 DG = -31.0 kJ mol^{-1}

Overall DG = +15.7 kJ mol^{-1} - 31.0 kJ mol^{-1} = -15.3 kJ mol^{-1}

DG = - RT ln K

- 15.3 kJ mol^{-1} = - (8,314 x 10^{-3} kJ K^{-1} mol^{-1})(310 K)ln K

ln K = 5.9

K = 380

5.37 $\ln K - \ln K' = -\frac{\Delta H}{R}\left(\frac{1}{T} - \frac{1}{T'}\right)$

$$4.14 = -\frac{\Delta H}{8.314\ \mathrm{J\ K^{-1}\ mol^{-1}}}\left(\frac{1}{323} - \frac{1}{373}\right)$$

DH = **-8.3×10^2 kJ mol^{-1}**

Additional problems

5.1 Hydrogen sulfide decomposes according to the equation
$2H_2S(g) \rightleftharpoons 2H_2(g) + S_2(g)$

When 4.0 mol H_2S were placed in a 1.0 L vessel and heated to 600° C the equilibrium mixture was found to contain 0.044 mol $S_2(g)$. What is the value of K for this temperature?

5.2 Write the expression for K for the vaporization of water.

5.3 What is the concentration of H_3O^+ in 0.010 M $Ca(OH)_2$?

5.4 K_a for HCN is 4.9×10^{-10}. What is K_b for CN^- ?

5.5 Calculate the degree of ionization of acetic acid in a 0.10 M aqueous solution.

5.6 Calculate the degree of ionization of acetic acid in the above solution when sufficient HCl has been added to make the solution 0.010 M HCl.

5.7 Calculate the pH of a solution in which 10.0 mL of 0.10 M NaOH is added to 5.0 mL of 0.10 M HCl.

5.8 Predict whether aqueous solutions of the following salts will be acidic, basic or neutral.
a) NH_4Cl
b) KCN
c) $Zn(NO_3)_2$

5.9 By experiment, it was found that 1.20×10^{-3} mol of lead iodide, PbI_2, will dissolve in 1.0 L of aqueous solution. What is K_s?

5.10 Will the solubility of $CaCO_3$ increase or decrease if a strong acid is added to the solution?

Chapter 6

Electrochemistry

6.1 $HgCl_2(s) \rightleftharpoons Hg^{2+}(aq) + 2Cl^-(aq)$ $K = [Hg^{2+}][Cl^-]^2$
$[Cl^-] = 2 \times [Hg^{2+}]$; therefore $K = 4[Hg^{2+}]^3$

and the solubility of the salt is

$$S = [Hg^{2+}] = \left(\frac{1}{4}K\right)^{1/3} M$$

From $\Delta G^\ominus = \Delta G_f^\ominus(Hg^{2+}) + 2\Delta G_f^\ominus(Cl^-) - \Delta G_f^\ominus(HgCl_2)$
$= +164.40 = 2 \times (-131.23) - (-178.6)$ kJ mol^{-1}
$= 80.54$ kJ mol^{-1}

$$\ln K = \frac{-\Delta G^\ominus}{RT} = \frac{-80.54 \times 10^3\ \mathrm{J\,mol^{-1}}}{8.314\ \mathrm{JK^{-1}mol^{-1}} \times 298.15\ \mathrm{K}} = -32.49$$

Therefore $K = 7.75 \times 10^{-15}$

and S = **1.25×10^{-5} mol L^{-1}**

6.2 $\Delta E = \frac{RT}{F}\ln\frac{a_1(H^+)}{a_2(H^+)}$ [Nernst equation]

$$= \frac{25.7\ \mathrm{mV}}{2}\ln\frac{Q_2}{Q_1} = \frac{25.7\ \mathrm{mV}}{2}\ln\frac{(0.025)^2}{(0.0050)^2}$$

= **41 mV**

6.3 $K_a = 8.4 \times 10^{-4} = \frac{x^2}{0.005 - x} \approx \frac{x^2}{0.005}$

$x = 2.05 \times 10^{-3}$

pH = 2.69

Similar calculation for 0.025 M, gives pH = 2.34

DE = 0.0592 pH_2 - 0.0592 pH_1 = **-0.021 mV**

6.4 R: $Cl_2(g) + 2e^- \rightarrow 2Cl^-(aq)$ $E^{\ominus} = +1.36V$

L: $Mn^{2+}(aq) + 2e^- \rightarrow Mn(s)$ $E^{\ominus} = ?$

The cell corresponding to these half reactions is

$Mn|MnCl_2(aq)|Cl_2(g)|Pt$ $E^{\ominus} = 1.36V - E^{\ominus}(Mn, Mn^{2+})$

Hence $E^{\ominus}(Mn, Mn^{2+}) = 1.36V - 2.54V =$ **-1.18V**

6.5

(a) R: $Ag^+(aq) + e^- \rightarrow Ag(s)$ +0.80 V

L: $Ag(s) \rightarrow Ag^+(aq) + e^-$ -0.80 V

Overall (R-L): $Ag^+(aq)$ (R) $\rightarrow$ $Ag^+(aq)$ (L)

(b) $H_2(g)$ (R) $\rightarrow$ $H_2(g)$(L)

(c) R: $MnO_2(s) + 4H^+(aq) + 2e^- \rightarrow Mn^{2+}(aq) + 2H_2O$

L: $3[Fe(CN)_6]^{3-}(aq) + 3e^- \rightarrow 3[Fe(CN)_6]^{4-}(aq)$

R - L: $3MnO_2(s) + 12H^+(aq) + 6[Fe(CN)_6]^{4-}(aq) \rightarrow$
$3Mn^{2+}(aq) + 6[Fe(CN)_6]^{3-}(aq)$

(d) R: $Br_2(l) + 2e^- \rightarrow 2Br^-(aq)$

L: $Cl_2(g) + 2e^- \rightarrow 2Cl^-(aq)$

Cell: R-L: $Br_2(l) + 2Cl^-(aq) \rightarrow Cl_2(g) + 2Br^-(aq)$

(e) R: $Sn^{4+}(aq) + 2e^- \rightarrow Sn^{2+}(aq)$

L: $2Fe^{3+}(aq) + 2e^- \rightarrow 2Fe^{2+}(aq)$

R - L: $Sn^{4+}(aq) + 2Fe^{2+}(aq) \rightarrow Sn^{2+}(aq) + 2Fe^{3+}(aq)$

(f) R: $MnO_2(s) + 4H^+(aq) + 2e^- \rightarrow Mn^{2+}(aq) + 2H_2O(l)$

L: $Fe^{2+}(aq) + 2e^- \rightarrow Fe(s)$

R - L: $Fe(s) + MnO_2(s) + 4H^+(aq)$
$\rightarrow Fe^{2+}(aq) + Mn^{2+}(aq) + 2H_2O(l)$

6.6 Nernst equations:

a) $$E^o_{cell} = -\frac{RT}{F}\ln\frac{m_L}{m_R}$$

b) $$E^o_{cell} = -\frac{RT}{F}\ln\frac{P_{H_2L}}{P_{H_2R}}$$

c) $$E^o_{cell} = -\frac{RT}{6F}\ln\frac{(Mn^{2+})^3\left(Fe(CN)_6^{3-}\right)^6}{\left(H^+\right)^{12}\left(Fe(CN)_6^{4-}\right)^6}$$

d) $$E^o_{cell} = -\frac{RT}{2F}\ln\frac{[Br^{2-}]^2 P_{Cl_2}}{\left[Cl^-\right]^2}$$

e) $$E^o_{cell} = -\frac{RT}{2F}\ln\frac{[Mn^{2+}]\left[Fe^{2+}\right]}{\left[H^+\right]^4}$$

6.7 First we identify the half-reactions, and then set up the corresponding cell:

(a)

R: $Pb^{2+}(aq) + 2e^- \rightarrow Pb(s)$ -0.13 V

L: $Fe^{2+}(aq) + 2e^- \rightarrow Fe(s)$ -0.44 V

Hence the cell is

$Fe(s)|FeSO_4(aq)||PbSO_4(aq)|Pb(s)$ **n = 2**

(b) R: $Hg_2Cl_2(s) + 2e^- \rightarrow 2Hg(l)$ +0.27

L: $2H^+(aq) + 2e^- \rightarrow H_2(g)$ 0

The cell is:
$Pt|H_2(g)|H^+(aq)|Hg_2Cl_2(s)|Hg(l)$ **n = 2**

(c) R: $O_2(g) + 4H^+(aq) + 4e^- \rightarrow 2H_2O(l)$ +1.23 V

L: $4H^+(aq) + 4e^- \rightarrow 2H_2(g)$ 0

The cell is:
$Pt|H_2(g)|H^+(aq),\ H_2O|O_2(g)|Pt$ **n = 4**

(d) R: $O_2(g) + 2H^+(aq) + 2e^- \rightarrow 2H_2O_2$ +0.695 V

L: $2H^+(aq) + 2e^- \rightarrow H_2(g)$ 0

The cell is **n = 2**

$Pt|H_2(g)|H^+(aq),\ H_2O_2(aq)|O_2(g)|Pt$

(e) R: $I_2(s) + 2e^- \rightarrow 2I^-$ +0.54 V

L: $2H^+(aq) + 2e^- \rightarrow H_2(g)$ 0

The cell is

$Pt|H_2(g)|H^+(aq), I^-(aq)|I_2(s)|Pt$

or more simply
$Pt|H_2(g)|HI(aq)|I_2(s)|Pt$ **n = 2**

(f) R: $Cu^{2+}(aq) + e^- \rightarrow Cu^+(aq)$ +0.520 V

L: $Cu^+(aq) + e^- \rightarrow Cu(s)$ +0.154 V

$CuCl_2(aq) | CuCl(aq)|Cu(s)$ **n = 1**

6.8 See the solutions above where we have used $E^\ominus = E^\ominus_R - E^\ominus_L$
a) **See Nernst equation**

b) **See Nernst equation**

c) **+0.87 V**

d) **-0.27 V**

e) **-0.62 V**

f) **+1.67 V**

6.9 See the solutions above where we have used $E^\ominus = E^\ominus_R - E^\ominus_L$

a) **+0.08 V**

b) **+0.27 V**

c) **+1.23 V**

d) **+0.695 V**

e) **+0.54 V**

f) **+0.366 V**

6.10 a) |E| increases

b) |E| increases

c) |E| decreases

d) |E| decreases

e) |E| increases

6.11 a) |E| increases

b) |E| increases

c) |E| decreases

d) |E| decreases

e) |E| increases

f) |E| no change

6.12 R: $2Tl^{+}(aq) + 2e^{-} \rightarrow 2Tl(s)$ -0.34 V

L: $Hg^{2+}(aq) + 2e^{-} \rightarrow Hg(l)$ +0.86 V

(a) Combining for the cell potential $E^{\ominus} = E^{\ominus}_{R} - E^{\ominus}_{L} = -1.20$ V

(b) Overall: $2Tl^{+}(aq) + Hg(l) \rightarrow 2Tl(s) + Hg^{2+}(aq)$

$$Q = \frac{a(Hg^{2+})}{a(Tl^{+})}, \nu = 2$$

$$E = E^{\ominus} - \frac{RT}{\nu F}\ln Q$$

$$= -1.20\,V - \frac{25.693\,mV}{2} \times \ln\frac{0.150}{(0.93)^2}$$

$=-1.20$ V + 0.023 V = **-1.18 V**

6.13 (a) $Ca(s) + 2H_2O(l) \rightarrow Ca(OH)_2(aq) + H_2(g)$

$E^{\ominus}$ = +2.04 V

Therefore $\Delta G^{\ominus} = -\nu FE^{\ominus}$

= -2 x 96.485 kC mol^{-1} x 2.04 V = **-394 kJ mol^{-1}**

(b) same as above

(c) $E^{\ominus}$ = -0.39 V

Therefore $\Delta G^{\ominus} = -\nu FE^{\ominus}$

= -2 x 96.485 kC mol^{-1} x (-0.39 V) = **+75 kJ mol^{-1}**

(d) $E^{\ominus}$ = +1.51 V

Therefore $\Delta G^{\ominus} = -\nu FE^{\ominus}$

= -2 x 96.485 kC mol^{-1} x 1.51 V = **-291 kJ mol^{-1}**

(e) same as above

(f) $E^{\ominus}$ = -2.58 V

Therefore $\Delta G^{\ominus} = -\nu FE^{\ominus}$

= -2 x 96.485 kC mol^{-1} x (-2.58 V) = **-498 kJ mol^{-1}**

6.14 (a) $E^{\ominus} = \frac{-\Delta G^{\ominus}}{\nu F} = \frac{+62.5\,kJ\,mol^{-1}}{2 \times 96.485\,kC\,mol^{-1}}$ = **+0.324 V**

(b) $E^{\ominus} = E^{\ominus}(Fe^{3+},Fe^{2+}) - E^{\ominus}(Ag,Ag_2CrO_4, CrO_4^{2-})$

Therefore,
$E^{\ominus}(Ag,Ag_2CrO_4, CrO_4^{2-}) = E^{\ominus}(Fe^{3+},Fe^{2+}) - E^{\ominus}$

$= +0.77 - 0.324 \text{ V} = \underline{\mathbf{+0.45 V}}$

6.15 $E = E^{o}_{cell} - \dfrac{0.0592}{n}\log Q$

$$= \frac{0.0592}{n}\log\frac{m_L}{m_R}$$

$= \underline{\mathbf{-0.023 V}}$

6.16 $2Ag(s) + Cu^{2+}(aq) \rightarrow 2Ag^{+}(aq) + Cu(s)$

$E = E^{\ominus}(Cu^{2+},Cu) - E^{\ominus}(Ag^{+}, Ag) = +0.34 - 0.80 \text{ V} = \underline{\mathbf{-0.46 V}}$

$\Delta G^{\ominus} = 2\Delta G^{\ominus}_f (Ag^{+}, aq) - \Delta G^{\ominus}_f (Cu^{2+}, aq)$

$= 2 \times 77.1 - (64.49) \text{ kJ mol}^{-1} = \underline{\mathbf{+89.7 kJ mol^{-1}}}$

$\Delta H^{\ominus} = 2\ \Delta H^{\ominus}_f (Ag^{+}, aq) - \Delta H^{\ominus}_f (Cu^{2+}, aq)$

$= 2 \times 105.58 - (64.77) \text{ kJ mol}^{-1} = \underline{\mathbf{146 kJ mol^{-1}}}$

$$\left(\frac{\partial \Delta G^{\ominus}}{\partial T}\right)_p = -\Delta S^{\ominus} = \frac{\Delta G^{\ominus} - \Delta H^{\ominus}}{T} \quad \text{[DG = DH - TDS]}$$

DS = 190 J mol^{-1} K^{-1}

Therefore $\Delta G^{\ominus}(308 \text{ K}) \approx 146 - (308) \times 0.190 \text{ kJ mol}^{-1}$

$\approx \underline{\mathbf{+87.9 kJ mol^{-1}}}$

6.17 a) $CO_3^{2-} + H_2 \rightarrow HCO_3^- + H^+$

$\Delta G^{\ominus} = \Delta G^{\ominus}(HCO_3^-) - \Delta G^{\ominus}(CO_3^{2-}) = -58.96 \text{ kJ mol}^{-1}$

$$E^o = -\frac{\Delta G}{nF} = \frac{58.96 \text{ kJ}}{2 \times 9.4865 \times 10^4} = \underline{\mathbf{0.3108\ V}}$$

b) Use the above equation for the reaction and

$2H_2O + 2e^- \rightarrow H_2 + 2OH^-$ -0.83 V

E_{cell} = R-L = -0.83 + 0.3108 V = **<u>-0.519 V</u>**

c) $\mathbf{E = E_{cell}} - \frac{RT}{2F}\ln\frac{[HCO_3^-][OH^-]}{[CO_3^{2-}]}$

d) E = E' - 2.303 RT pH = E' - 59.160 mV x 7

e) DG^o = -RT ln K

pK = **<u>10.3</u>**

6.18 $Cu_3(PO_4)_2(s) \rightleftharpoons 3Cu^{2+}(aq) + 2PO_4^{3-}(aq)$

$K_s = a(Cu^{2+})^3 a(PO_4^{3-})^2 \approx mCu^{2+})^3 mPO_4^{3-})^2$

(a) $S = m(Cu_3(PO_4)_2) = \frac{1}{3}m(Cu^{2+})$

However, $m(PO_4^{3-}) = \frac{2}{3}m(Cu^{2+})$

Therefore

$K_s = \frac{4}{9}m(Cu^{2+})^5$, which implies that $S = \frac{1}{3} \times (\frac{4}{9}K_{sp})^{1/5}$

$$S = \frac{1}{3} \times (\frac{4}{9} \times 1.3 \times 10^{-37})^{1/5} = \underline{\mathbf{1.6 \times 10^{-8}\ [mol\ kg^{-1}]}}$$

(b) The cell reaction is

R: $Cu^{2+}(aq) + 2e^- \rightarrow Cu(s)$ +0.34 V
L: $2H^+(aq) + 2e^- \rightarrow H_2(g)$ 0

Overall: $Cu^{2+}(aq) + H_2(g) \rightarrow Cu(s) + 2H^+(aq)$ +0.34 V
Using the Nernst equation

$$E = E^{\ominus} - \frac{RT}{\nu F} \ln Q$$

$$= 0.34\ \text{V} - \frac{25.693 \times 10^{-3}\ \text{V}}{2} \ln \frac{a(H^+)^2}{a(Cu^+)}$$

Since $m(Cu^{2+}) = 3S$, the following substitution can be made:

$$E = 0.34\ \text{V} - \frac{25.693 \times 10^{-3}\ \text{V}}{2} \ln \frac{1}{3 \times (1.6 \times 10^{-8})}$$
$$= 0.34\ \text{V} - 0.22\ \text{V} = \underline{\mathbf{+0.12\ V}}$$

6.19 (a) $Sn(s) + Sn^{4+}(aq) \rightleftharpoons 2Sn^{2+}(aq)$

R: $Sn^{4+} + 2e^- \rightarrow Sn^{2+}(aq)$ +0.15 V

L: $Sn^{2+}(aq) \rightarrow Sn(s)$ -0.14 V

$E^{\ominus} = +0.29$ V

$$\ln K = \frac{\nu F E^{\ominus}}{RT} = \frac{2 \times 0.29\ \text{V}}{25.693\ \text{mV}} = +22.6,\ K = \underline{\mathbf{6.5 \times 10^9}}$$

(b) $Sn(s) + 2AgBr(s) \rightleftharpoons SnBr_2(aq) + 2Ag(s)$
R: $2AgBr(s) + 2e^- \rightarrow 2Ag(s) + 2Br^-(aq)$ +0.07 V

L: $Sn^{2+}(aq) \rightarrow Sn(s)$ -0.14 V

$E^{\ominus}$ = +0.21 V

$$\ln K = \frac{2 \times 0.21\ \text{V}}{25.693} \times 10^3 = +16.3,\ K = \underline{\mathbf{1.2 \times 10^7}}$$

(c) $Fe(s) + Hg(NO_3)_2(aq) \rightleftharpoons Hg(l) + Fe(NO_3)_2(aq)$

R: $Hg^{2+}(aq) + 2e^- \rightarrow Hg(l)$ +1.62 V
L: $Fe^{2+}(aq) + 2e^- \rightarrow Fe(s)$ -0.44 V

$E^{\ominus}$ = +2.06 V

$$\ln K = \frac{2 \times 2.06\ \text{V}}{25.693} \times 10^3 = 160,\ K = \underline{\mathbf{4.4 \times 10^{69}}}$$

(d) $Cd(s) + CuSO_4(aq) \rightleftharpoons Cu(s) + CdSO_4(aq)$

R: $Cu^{2+}(aq) + 2e^- \rightarrow Cu(s)$ +0.34 V
L: $Cd^{2+}(aq) \rightarrow Cd(s)$ -0.40 V

$E^{\ominus}$ = +0.74 V

$\ln K = 57.6$, $K = \underline{\mathbf{1.0 \times 10^{25}}}$

(e) $Cu^{2+}(aq) + Cu(s) \rightleftharpoons 2Cu^{+}(aq)$

R: $Cu^{2+}(aq) + e^- \rightarrow Cu^{+}(aq)$ +0.16 V
L: $Cu^{+}(aq) + e^- \rightarrow Cu(s)$ +0.52 V

$E^{\ominus}$ = -0.36 V

$$\ln K = \frac{-0.36\ \text{V}}{25.693\ \text{mV}} = -14.0,\ K = \underline{\mathbf{8.2 \times 10^{-7}}}$$

6.20 $S(AgCl) = m(Ag^+)$

$AgCl(s) \rightleftharpoons Ag^+(aq) + Cl^-(aq)$ $K_s \approx m(Ag^+)m(Cl^-)/m^{\ominus 2}$

Since $m(Ag^+) = m(Cl^-)$

$K_s \approx m(Ag^+)^2/m^{\ominus 2} = S^2/m^{\ominus 2} = (1.34 \times 10^{-5})^2$

$\approx$ **1.80 x 10^{-10}**

$S(BaSO_4) = m(Ba^+)$

$BaSO_4(s) \rightleftharpoons Ba^{2+}(aq) + SO_4^{2-}(aq)$

As above, $K_s \approx S^2/\, m^{\ominus 2} = (9.51 \times 10^{-4})^2 =$ **9.04 x 10^{-7}**

6.21 The half-reaction is

$Cr_2O_7^{2-}(aq) + 14H^+(aq) + 6\,e^- \rightarrow 2Cr^{3+}(aq) + 7H_2O(l)$

The reaction quotient is

$$Q = \frac{a(Cr^{3+})^2}{a(Cr_2O_7^{2-})a(H^+)^{14}} \qquad \nu = 6$$

Hence,

$$E = \mathbf{E^{\ominus}} - \frac{\mathbf{RT}}{\mathbf{6F}}\,\mathbf{ln}\,\frac{\mathbf{a(Cr^{3+})^2}}{\mathbf{a(Cr_2O_7^{2-})a(H^+)^{14}}}$$

6.22 R: $2AgCl(s) + 2e^- \rightarrow 2Ag(s) + 2Cl^-(aq)$ +0.22 V
L: $2H^+(aq) + 2e^- \rightarrow H_2(g)$ 0

Overall: $2AgCl(s) + H_2(g) \rightarrow 2Ag(s) + 2Cl^-(aq) + 2H^+(aq)$

$Q = a(H^+)^2 a(Cl^-)^2 \quad n = 2$

$= a(H^+)^4 \qquad [a(H^+) = a(Cl^-)]$

Therefore from the Nernst equation,

$$E = E^{\ominus} - \frac{RT}{2F}\ln a(H^+)^4 = E^{\ominus} - \frac{2RT}{F}\ln a(H^+)$$

$$= E^{\ominus} + 2 \times 2.303\frac{2RT}{F}\,pH$$

Rearranging and substituting in:

$$pH = \frac{F}{2 x 2.303RT} x (E - E^{\ominus}) = \frac{E - 0.22\ V}{0.1183\ V}$$

$= \mathbf{\underline{0.78}}$

6.23 R: $AgBr(s) + e^- \rightarrow Ag(s) + Br^-(aq)$

L: $Ag^+(aq) + e^- \rightarrow Ag(s)$

Overall: $AgBr(s) \rightarrow Br^-(aq) + Ag^+(aq)$

Therefore since the cell reaction is the solubility equilibrium, for a saturated solution there is no further tendency to dissolve and so **E = 0**.

6.24 R: $Ag^+(aq) + e^- \rightarrow Ag(s)$ +0.80 V

L: $AgI(s) + e^- \rightarrow Ag(s) + I^-(aq)$ -0.15 V

+0.9509 V

Overall: $Ag^+(aq) + I^-(aq) \rightarrow AgI(s)$ n = 1

$$\ln K = \frac{0.9509\ V}{25.693\ mV} = 37.010,\ K = 1.184 \times 10^{16}$$

However, $K_s = K^{-1}$ since the solubility equilibrium is written as the reverse of the cell reaction. Therefore K_s = **8.5 x 10^-17^**.

The solubility is obtained from $m(Ag^+) \approx m(I)$ and $S = m(Ag^+)$ $K_s \approx m(Ag^+)^2$, so that

$S \approx (K_s)^{1/2}\ mol\ L^{-1} = (8.45 \times 10^{-17})^{1/2}\ mol\ L^{-1}$
$= \mathbf{9.19 \times 10^{-9}\ mol\ L^{-1}}$

6.25 R: $HgSO_4(s) + 2e^- \rightarrow 2Hg(l) + SO_4^{2-}(aq)$ +0.62 V

L: $PbSO_4(s) + 2e^- \rightarrow 2Pb(s) + SO_4^{2-}(aq)$ -0.36 V
R - L: $HgSO_4(s) + 2Pb(s) \rightarrow 2Hg(l) + PbSO_4(s)$

A suitable cell would be
$Pb(s)|PbSO_4(s)|H_2SO_4(aq)|Hg_2SO_4(s)|Hg(l)$

The electrode potentials are

$$E_R = E_R^\ominus - \frac{RT}{2F}\ln a(SO_4^{2-}) = E_R^\ominus - \frac{RT}{2F}\ln(K_S(Hg_2SO_4))^{1/2}$$

$$E_L = E_L^\ominus - \frac{RT}{2F}\ln a(SO_4^{2-}) = E_L^\ominus - \frac{RT}{2F}\ln(K_S(PbSO_4))^{1/2}$$

because $K_S = a_+a_- = (a_-)^2$

Therefore,

$$E = 0.98\text{ V} - \frac{RT}{2F}\ln\frac{K_S(Hg_2SO_4)^{1/2}}{K_S(PbSO_4)^{1/2}}$$

$$E = 0.98\text{ V} - \frac{RT}{4F}\ln\frac{K_S(Hg_2SO_4)}{K_S(PbSO_4)}$$

$$= 0.98\text{ V} - \frac{25.693\text{ mV}}{4}\ln\frac{6.6\times10^{-7}}{1.6\times10^{-8}}$$

= 0.98 V - 0.026 V = **0.95 V**

6.26 (a) $H_2(g) + 1/2\ O_2(g) \rightarrow H_2O(l)$

$\Delta G^\ominus = \Delta_f G^\ominus(H_2O, l) = -237.13\text{ kJ mol}^{-1}$

$$E^\ominus = -\frac{\Delta G^\ominus}{\nu F} = \frac{+237.13\text{ kJ mol}^{-1}}{2\times(96.485\text{ kC mol}^{-1})} = \mathbf{+1.23\ V}$$

(b) $C_6H_6(g) + 7\ 1/2\ O_2(g) \rightarrow 6CO_2(g) + 3H_2O(l)$

$$\Delta G^\ominus = 6\Delta G_f^\ominus(CO_2, g) + 3\Delta G_f^\ominus(H_2O, l) - \Delta G_f^\ominus(C_6H_6, g)$$

$$= 6 \times (-394.36) + 3 \times (-237.13) - (+124.3) \text{ kJ mol}^{-1}$$
$$= -3202 \text{ kJ mol}^{-1}$$

To find the number of electrons transferred, note that the cathode half-reaction is the reduction of oxygen to produce $3H_2O$:

$3/2\ O_2(g) + 6e^- + 6H^+(aq) \rightarrow 3H_2O(l)$ $\quad n = 6$
Therefore,

$$E^o = -\frac{\Delta G}{nF} = \frac{-3202 \text{ kJ mol}^{-1}}{6 \times 96.485} = \underline{\mathbf{+5.53\ V}}$$

6.27 R: $CH_4 + 3/2\ O_2 \rightarrow CO + 2H_2O$
L: $CH_4 + 2O_2 \rightarrow CO_2 + 2H_2O$

For R:

$$\Delta G^\ominus = \Delta G_f^\ominus(CO, g) + 2\Delta G_f^\ominus(H_2O, l) - \Delta G_f^\ominus(CH_4, g)$$

$$= (-137.17) + 2 \times (-237.13) - (-50.72) \text{ kJ mol}^{-1}$$

$$= -560.7 \text{ kJ mol}^{-1}$$

$$E = -\frac{\Delta G}{nF} = \frac{560.7 \text{ kJ mol}^{-1}}{4 \times 96.485} = +1.45 \text{ V}$$

For L:

$$\Delta G^\ominus = \Delta G_f^\ominus(CO_2, g) + 2\Delta G_f^\ominus(H_2O, l) - \Delta G_f^\ominus(CH_4, g)$$
$$= (-394.36) + 2 \times (-237.13) - (-50.72) \text{ kJ mol}^{-1}$$
$$= -817.9 \text{ kJ mol}^{-1}$$

$E = +2.12$ V

<u>Cathode will be left side.</u>

b) $E^o = 2.12 - 1.45 =$ **0.67 V**

6.28 $MnO_4^- + 8H^+ + 5e^- \rightarrow Mn^{2+} + 4H_2O$ $\qquad E^\ominus = 1.51$ V

Using the Nernst equation to determine the reduction potential for pH 7.00

$$E = E^\ominus - \frac{25.7\text{ mV}}{\nu}\ln Q$$

$$\ln Q = \ln\frac{1}{a(H^+)^8} = 8 \times 2.303 \times pH$$

$$E = 1.51\text{ V} - \frac{25.7\text{ mV} \times 8 \times 2.303 \times 6.00}{5} = \underline{\mathbf{+0.94\ V}}$$

6.29 Adding up the overall cell potential gives a biological reduction potential for the cell of -0.22 V.

$$\ln K = \frac{\nu F E^\ominus}{RT} = \frac{2 \times (-0.22)}{25.69\text{ mV}} = -17.1$$

$K =$ **3.6 x 10^-8** ($K = \underline{\mathbf{3.6 \times 10^{-8}}}$)

6.30 $CH_3COCOOH + 2H^+ + 2e^- \rightleftharpoons CH_3CH(OH)COOH$

$$E^{\ominus\prime} = E^\ominus - \frac{RT}{2F}\ln Q \quad \text{and} \quad Q = \frac{1}{a(H^+)^2}$$

$$E^{\ominus\prime} = E^\ominus - \frac{2.303\,RT}{F} \times pH$$

$$= E^\ominus - (59.2\text{ mV}) \times pH$$

At biological conditions the pH = 7.0 Therefore,

$E^\ominus = -0.19\text{ V} + (0.0592)(7.0)\text{ V} =$ **+0.22 V**

Additional problems

6.1 A cell is based on the reaction
$MnO_4^-(aq) + H^+(aq) + ClO_3^-(aq) \rightleftharpoons ClO_4^-(aq) + Mn^{2+}(aq) + H_2O(l)$
What are the half reactions for the electrodes and the standard cell potential?

6.2 Calculate DG for the reaction
$Ag^+(aq) + Fe^{2+}(aq) \rightarrow Fe^{3+}(aq) + Ag(s)$

6.3 For the cell reaction
$Mn(s) + Cl_2(g) \rightarrow MnCl_2(aq) \quad E^\ominus = 2.54\ V$
predict whether E is larger or smaller than $E^\ominus$ under the following cases:

a) $[Mn^{2+}] = 2.0\ M$, $[Cl_2] = 1.0\ M$

b) $[Cl_2] = 2.0\ M$, $[Mn^{2+}] = 1.0\ M$

6.4 Calculate DG and K for the reaction
$2\ Na(s) + 2H_2O(l) \rightarrow 2NaOH\ (aq) + H_2(g)$

6.5 Calculate the standard cell potential for the following reaction from standard free energies of formation
$Zn(s) + Cl_2(g) \rightarrow Zn^{2+}(aq) + 2Cl^-(aq)$

6.6 What is the effect of pH on the potential of the following cell
$Cd(s)|Cd^{2+}||\ H^+|\ H_2(g)|\ Pt(s)$?

6.7 Calculate the cell potential for the cell corresponding to the reaction
$Zn(s) + Cu^{2+}(aq) \rightarrow Zn^{2+}(aq) + Cu(s)$
when $[Zn^{2+}] = 0.01\ M$, $[Cu^{2+}] = 2.0\ M$

6.8 What is the pH of a solution whose cell potential is +0.0592 V when measured with a standard hydrogen electrode?

6.9 What is the zinc ion concentration in the cell
$Zn|Zn^{2+}||\ Cu^{2+}(10.0\ M)|Cu(s)$ if the cell potential is 1.30 V?

6.10 What is the effect of acid rain on the corrosion of iron?
[Use the equation $Fe(s) \rightarrow Fe^{2+}(aq)$]

6.11 Calculate the equilibrium constant for the reaction
$Zn(s) + 2H^{+}(aq) \rightleftharpoons Zn^{2+}(aq) + H_2(g)$

6.12 Is the reaction
$Zn(s) + Cl_2(g) \rightarrow Zn^{2+}(aq) + 2Cl^{-}(aq)$
spontaneous under standard conditions?

Chapter 7

The rates of reactions

7.1 Since the rate of formation of C is known, the reaction stoichiometry can be used to determine the rates of consumption and formation of the other participants in the reaction.

[A] = 2/3[C] = **$1.5\ mol\ L^{-1}\ s^{-1}$**

[B] = 1/3[C] = **$0.73\ mol\ L^{-1}\ s^{-1}$**

[D] = 2/3[C] = **$1.5\ mol\ L^{-1}\ s^{-1}$**

7.2 The rate has units of concentration per unit time ($mol\ L^{-1}\ s^{-1}$)

rate ($mol\ L^{-1}\ s^{-1}$) = k [A in $mol\ L^{-1}$][B in $mol\ L^{-1}$][C in $mol\ L^{-1}$]

k must be in units of $\dfrac{1}{(mol\ L^{-1})^2} \times s^{-1}$ or **$mol^{-2}\ L^2\ s^{-1}$**

7.3 a) For a second order reaction, denoting units of k by [k]:

$M\ s^{-1} = [k] \times M^2$, implying the [k] =**$(molecules\ m^{-3})^{-1} s^{-1}$**

b) [k] = **$kPa^{-1}s^{-1}$**

For a third order reaction:

$M\ s^{-1} = [k] \times M^3$, implying [k] = **$(molecules\ m^{-3})^{-2}\ s^{-1}$**

b) For a third order reaction:

$kPa\ s^{-1} = [k] \times kPa^3$, implying [k] = **$kPa^{-2}s^{-1}$**

7.4 rate = $k[N_2O_5]$

$kt_{1/2} = \ln 2 \qquad t_{1/2} = \dfrac{\ln 2}{(3.38 \times 10^{-5}\ s^{-1})}$

$t_{1/2} =$ **$\underline{2.05 \times 10^4}$**

a) After 10 s very little N_2O_5 will have decomposed (note the halflife).

$p(N_2O_5) = p_o(N_2O_5)\ e^{-kt}$

a) $p(N_2O_5) = 500\ \text{Torr} \times e^{-(0.0000338) \times (10)} =$ **$\underline{500\ \text{Torr}}$**

b) $p(N_2O_5) = 500\ \text{Torr} \times e^{-(0.0000338) \times (600)} =$ **$\underline{490\ \text{Torr}}$**

7.5 $\ln \dfrac{[A_o]}{[A]} = kt$

$\ln \dfrac{\ln \dfrac{22.0}{56.0}}{1.22 \times 10^4\ s} = k =$ **$\underline{1.12 \times 10^{-4}\ s^{-1}}$**

7.6 $\ln \dfrac{[P_o]}{[P]} = kt$

$\ln P = kt$

$\ln (100\ Pa) = k(522\ s)$

$k =$ **$\underline{8.8 \times 10^{-3}\ s^{-1}}$**

7.7 $\dfrac{1}{[A]_t} - \dfrac{1}{[A]_0} = kt$

$\dfrac{1}{56.0} - \dfrac{1}{220} = k(1.22 \times 10^4\ s)$

$k = \underline{\mathbf{1.09 \times 10^{-6}\ mmol^{-1} s^{-1}}}$

7.8 $\ln \dfrac{[A_o]}{[A]} = kt$

$$\dfrac{\ln \dfrac{22.0}{56.0}}{1.22 \times 10^4\ s} = k = \underline{\mathbf{1.12 \times 10^{-4}\ s^{-1}}}$$

7.9 $\dfrac{1}{[P]_t} - \dfrac{1}{[P]_0} = kt$

$$\dfrac{1}{100\ Pa \times 522\ s} = k = \underline{\mathbf{1.9 \times 10^{-5}\ Pa^{-1}\ s^{-1}}}$$

7.10 $(P_o - P_t)/t = \text{rate}$

$(21\ kPa - 10\ kPa)/770\ s = \underline{\mathbf{0.014\ kPa\ s^{-1}}}$

b) rate = P_o/t

$t = 21\ kPa/0.014\ kPa\ s^{-1} = \underline{\mathbf{1500\ s}}$

7.11 At $t_{1/2}$ $[A] = 1/2[A]_o$

$(1/2)^n = 1/64$ $n = 6$

$t = t_{1/2} \times 6 = 221\ s \times 6 = \underline{\mathbf{1330\ s}}$

7.12 $[^{14}C] = [^{14}C]_o\, e^{-kt}$, $k = \dfrac{\ln 2}{t_{1/2}}$

$$t = \frac{1}{k}\ln\frac{[^{14}C]_0}{[^{14}C]} = \frac{t_{1/2}}{\ln 2} \times \ln\frac{[^{14}C]_0}{[^{14}C]}$$

$$= \frac{5730\ y}{\ln 2} \times \ln\frac{1.00}{0.69} = \underline{\mathbf{3067\ y}}$$

7.13 $[^{90}Sr] = [^{90}Sr]_0 e^{-kt}$, $k = \frac{\ln 2}{t_{1/2}}$

$$k = \frac{\ln 2}{28.1\,y} = 0.0247\ y^{-1}$$

With [Sr] replaced by mass,

(a) $m = 1.00\ mg \times e^{-(0.0247) \times (19)} = \underline{\mathbf{0.63\ mg}}$

(b) $m = 1.00\ mg \times e^{-(0.0247) \times (75)} = \underline{\mathbf{0.16\ mg}}$

7.14 $kt = \frac{1}{[B]_0 - [A]_0}\ln\frac{[A]_0([B]_0 - x)}{([A]_0 - x)[B]_0}$

which rearranges to

$$x = \frac{[A]_0[B]_0\{e^{k([B]_0-[A]_0)t} - 1\}}{[B]_0\{e^{k([B]_0-[A]_0)kt} - [A]_0\}}$$

$$= \frac{(0.055)(0.150)\{e^{0.11(0.095)(15s)} - 1\}}{0.150\{e^{0.11(0.95)(15s)} - 0.055\}}$$

$= 8.37 \times 10^{-3}$ M

$[CH_3COOC_2H_5] = 0.150\ M - 0.0084\ M = \underline{\mathbf{0.142\ M}}$

b) After 15 min, or 900 s

x = 0.055 M

$[CH_3COOC_2H_5] = 0.150\ M - 0.055\ M =$ **0.095 M**

7.15 The amount of A in a second order reaction is

$$\frac{1}{[A]} - \frac{1}{[A]_0} = kt$$

Therefore

$$t = \frac{1}{k}\left\{\frac{1}{[A]_t} - \frac{1}{[A]_0}\right\}$$

$$= \frac{1}{1.24 \times 10^{-3}\ mol^{-1}\ L\,s^{-1}} \times \left\{\frac{1}{0.026\ mol\,L^{-1}} - \frac{1}{0.260\ mol\ L^{-1}}\right\}$$

$= \mathbf{2.79 \times 10^4\ s}$

7.16 $[B]_\infty = 1/2\ [A]_o$, hence $[A]_o = 0.624\ mol\ L^{-1}$. For the reaction $2A \rightarrow B$, $[A] = [A]_o - 2[B]$. Construct the following table

t/s	0	600	1200	1800	2400
$[B]/mol\ L^{-1}$	0	0.089	0.153	0.200	0.230
$[A]/mol\ L^{-1}$	0.624	0.446	0.318	0.224	0.164

The data are plotted. We see that the half-life of A from its initial concentration is approximately 1200 s, and that its half-life from the concentration at 1200 s is also 1200 s. This indicates a **first-order reaction.** This can be confirmed by plotting the data accordingly.

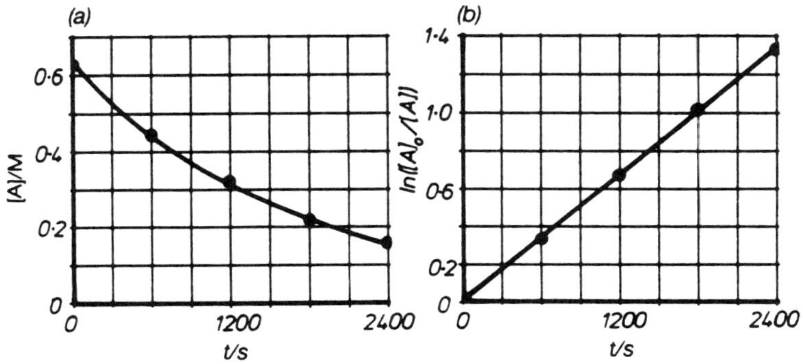

$\ln\frac{[A]_0}{[A]} = kt$ Draw up the following table:

t/s	0	600	1200	1800	2400
$\ln\frac{[A]_0}{[A]}$	0	0.34	0.67	1.02	1.34

and plot the points. The points lie on a straight line, which confirms first-order kinetics. The slope of the line = k_A = **5.6×10^{-4} s^{-1}.**

7.17 $\ln k = \ln A - \frac{E_a}{RT}$

$\ln k' = \ln A - \frac{E_a}{RT'}$

$$E_a = \frac{R \ln (k'/k)}{\left(\frac{1}{T} - \frac{1}{T'}\right)} = \frac{8.314 \text{ J K}^{-1} \text{ mol}^{-1} \times \ln\left(\frac{1.38 \times 10^{-3}}{1.78 \times 10^{-4}}\right)}{\frac{1}{292 \text{ K}} - \frac{1}{310 \text{ K}}}$$

= **85.6 kJ mol^{-1}**

For A, use

$A = k \times e^{E_a/RT}$

$= 1.78 \times 10^{-4} \text{ mol L}^{-1} \text{ s}^{-1} \times e^{85600/8.314 \times 292}$

= **3.66×10^{11} mol L^{-1} s^{-1}**

7.18 $E_a = \frac{R \ln (k'/k)}{\left(\frac{1}{T} - \frac{1}{T'}\right)}$

$$= \frac{8.314\ \mathrm{J\ K^{-1}\ mol^{-1}} \times \ln\left(\frac{K'}{1.1K'}\right)}{\frac{1}{T} - \frac{1}{298\ \mathrm{K}}} = 99.1\ \mathrm{kJ\ mol^{-1}}$$

T = **<u>299 K</u>**

7.19 One with larger E_a, 52 kJ mol^{-1}.

7.20 $E_a = \dfrac{R \ln (k'/k)}{\left(\frac{1}{T} - \frac{1}{T'}\right)}$

K' = 1.23 K

$$E_a = \frac{8.314\ \mathrm{J\ K^{-1}\ mol^{-1}} \times \ln(1.23)}{\frac{1}{293\ \mathrm{K}} - \frac{1}{300\ \mathrm{K}}}$$

= **<u>21.6 kJ mol^{-1}</u>**

7.21 $E_a = \dfrac{R \ln (k'/k)}{\left(\frac{1}{T} - \frac{1}{T'}\right)}$

$$= \frac{8.314\ \mathrm{J\ K^{-1}\ mol^{-1}} \times \ln(40)}{\frac{1}{277\ \mathrm{K}} - \frac{1}{298\ \mathrm{K}}} = \underline{\mathbf{120\ kJ\ mol^{-1}}}$$

7.22 <u>E_a = -21.6 kJ mol^{-1}</u>

7.23 $E_a = \dfrac{R \ln (k'/k)}{\left(\dfrac{1}{T} - \dfrac{1}{T'}\right)}$

$k = \ln \dfrac{2}{t_{1/2}}$ and $k' = \ln \dfrac{2}{2t_{1/2}}$

$k'/k = 1/2$

$$E_a = \frac{8.314 \text{ J K}^{-1} \text{ mol}^{-1} \text{ x } \ln(1/2)}{\dfrac{1}{293 \text{ K}} - \dfrac{1}{283 \text{ K}}} = \underline{\mathbf{48 \text{ kJ mol}^{-1}}}$$

7.24 $\ln k/k' = \dfrac{E_a}{R}\left(\dfrac{1}{T'} - \dfrac{1}{T}\right)$

$$= \frac{2.51 \text{ x } 10^5 \text{ J K}^{-1} \text{ mol}^{-1}}{8.314 \text{ J K}^{-1} \text{ mol}^{-1}}\left(\frac{1}{823} - \frac{1}{728}\right) = -4.787$$

$k/k' = t_{1/2}'/t = 8.34 \text{ x } 10^{-3}$

$t_{1/2}' = 8.34 \text{ x } 10^{-3} \text{ x } 6.5 \text{ x } 10^6 \text{ s} =$ **<u>5.42 x 10⁴ s</u>**

7.25 The first step is rate determining; hence
$v = k[H_2O_2][Br^-]$

The reaction is **first-order** in H_2O_2 and in Br^-, and **second-order overall.**

7.26 We assume a pre-equilibrium (that step is fast), and write

$K = \dfrac{[A]^2}{[A_2]}$ so that $[A] = K^{1/2}[A_2]^{1/2}$

The rate determining step then gives

$v = k_2[A][B] = \mathbf{k_2K^{1/2}[A_2]^{1/2}[B]}$

7.27 We assume a pre-equilibrium and write

$K = \dfrac{[\text{Unstable helix}]}{[A][B]}$, impling that [Unstable helix] = K[A][B]

The rate-determining step then gives

$v = k_1[\text{unstable helix}] = \mathbf{k_1K[A][B]}$

The equilibrium constant is the outcome of two processes

$$A + B \underset{k_{2'}}{\overset{k_2}{\rightleftharpoons}} [\text{Unstable helix}] \qquad K = \frac{k_2}{k_2'}$$

Therefore, $v = k[A][B]$ $\quad \mathbf{k = k_1k_2/k_2'}$

7.28 Maximum velocity = $k_b[E]_o$

Also $k = \dfrac{k_b[S]}{K_M + [S]}$

Therefore

$v = \dfrac{k_b[S][E]_0}{K_M + [S]}$ rearranging

$$k_b[E]_o = \left\{\frac{K_M + [S]}{[S]}\right\} v$$

$$= \left\{\frac{0.045 \text{ mol L}^{-1} + 0.110 \text{ mol L}^{-1}}{0.110 \text{ mol L}^{-1}}\right\} \text{x } (1.15 \text{ x } 10^{-3} \text{ mol L}^{-1} \text{ s}^{-1})$$

$= \underline{\mathbf{1.62 \text{ x } 10^{-3} \text{ M s}^{-1}}}$

7.29 From exercise 7.14 it follows that

$$\frac{[S]}{K_M+[S]} = 1/2$$ which is satisfied when **[S] = K_M**

7.30 $$\frac{d[R]}{dt} = 2\,k_1[R_2] - k_2[R][R_2] + k_3[R'] - 2\,k_4[R]^2$$

$$\frac{d[R']}{dt} = k_2[R][R_2\} - k_3[R']$$

Apply the steady-state approximation to both equations:

$2\,k_1[R_2] - k_2[R][R_2] + k_3[R'] - 2\,k_4[R]^2 = 0$

$k_2[R][R_2\} - k_3[R'] = 0$

The second solves to

$$[R'] = \frac{k_2}{k_3}[R][R_2]$$

and then the first solves to

$$[R] = \left(\frac{k_1}{k_4}[R_2]\right)^{1/2}$$

Therefore,

$$\frac{d[R_2]}{dt} = -k_1[R_2] - k_2[R_2][R] = \mathbf{-k_1[R_2] - k_2\left(\frac{k_1}{k_4}\right)^{1/2}[R_2]^{3/2}}$$

7.31 At 700 K, the branching explosion does not occur. At 800 K, it occurs beteen **0.16 kPa and 4.0 kPa**. At 900 K, branding occurs for pressures in excess of **0.11 kPa**.

7.32 Number of photons absorbed = Φ^{-1} x Number of molecules that react. Therefore,

$$\text{Number absorbed} = \frac{(1.075 \times 10^{-3}\ \text{mol}) \times (6.022 \times 10^{23}\ \text{einstein}^{-1})}{2.1 \times 10^{2}\ \text{mol einstein}^{-1}}$$

$= \underline{\mathbf{3.1 \times 10^{18}}}$

7.33 For a source of power P and wavelength λ, the amount of photons (n_γ) generated in a time t is

$$n_\gamma = \frac{Pt}{h\nu N_A} = \frac{P\lambda t}{hcN_A}$$

$$= \frac{(100\ \text{W}) \times (45) \times (60\ \text{s}) \times (490 \times 10^{-9}\ \text{m})}{(6.626 \times 10^{-34}\ \text{J s}) \times (2.998 \times 10^{8}\ \text{m s}^{-1}) \times (6.022 \times 10^{23}\ \text{mol}^{-1})}$$

$= 1.11$ mol

The amount of photons absorbed is 60 percent of this incident flux, or 0.660 mol. Therefore,

$$\Phi = \frac{0.297\ \text{mol}}{0.65 \times 1.11\ \text{mol}} = \underline{\mathbf{0.412}}$$

7.34 $\frac{d[A^-]}{dt} = k_1[AH][B] - k_2[A^-][BH^+] - k_3[A^-][A] = 0$

Therefore,

$$[A^-] = \frac{k_1[AH][B]}{k_2[BH^+] + k_3[A]}$$

and the rate of formation of product is

$$\frac{d[P]}{dt} = k_3[A][A^-] = \frac{k_1k_3[A][AH][B]}{k_2[BH^+] + k_3[A]}$$

7.35 $\frac{d[AH]}{dt} = k_3[HAH^+][B]$ [rate-determining]

$$K = \frac{[HAH^+]}{[HA][H^+]} \qquad \text{[pre-equilibrium]}$$

and hence

$$\frac{d[AH]}{dt} = k_3K[HA][H^+][B]$$

The acidity constant of the conjugate acid of B is

$$BH^+ + H_2O \rightleftharpoons B + H_3O^+, \quad K_a = \frac{[B][H^+]}{[BH^+]}$$

Therefore

$$\frac{d[AH]}{dt} = k_3KK_a[HA][BH^+]$$

7.36 Step 1: initiation [radicals formed]; Steps 2 and 3: propagation [new radicals formed]; Step 4: termination [non-radical products formed].

$$\frac{d[AH]}{dt} = -k_a[AH] - k_c[AH][B]$$

(i) $\dfrac{d[A]}{dt} = k_a[AH] - k_b[A] + k_c[AH][B] - k_d[A][B] \approx 0$

(ii) $\dfrac{d[B]}{dt} = k_b[A] - k_c[AH][B] - k_d[A][B] \approx 0$

(i + ii) $[A][B] = \left(\dfrac{k_c}{2k_d}\right)[AH]$

(i - ii) $[A] = \left(\dfrac{k_a + 2k_c[B]}{2k_b}\right)[AH]$

Solving for [A]:

$$[A] = k[AH], \quad k = \left(\frac{k_a}{4k_b}\right)\left\{1 + \left[1 + \frac{4k_bk_c}{k_ak_d}\right]^{1/2}\right\}$$

from which it follows that

$$[B] = \frac{k_a[AH]}{2k_d[A]} = \frac{k_a}{2kk_d}$$

and hence

$$\frac{d[AH]}{dt} = -k_a[AH] - \left(\frac{k_a k_c}{2kk_d}\right)[AH] = k_{eff}[AH]$$

with $k_{eff} = k_a + \left(\frac{k_a k_c}{2kk_d}\right)$

7.37 Construct a table for 1/S and $1/\nu$

$1/S[10^3]$	1.0	0.50	0.33	0.25	0.20
$1/\nu[10^6]$	0.90	0.56	0.44	0.38	0.34

A plot of these values will yeild a line with slope of K_M/ν_{max} and an intercept of $1/\nu_{max}$

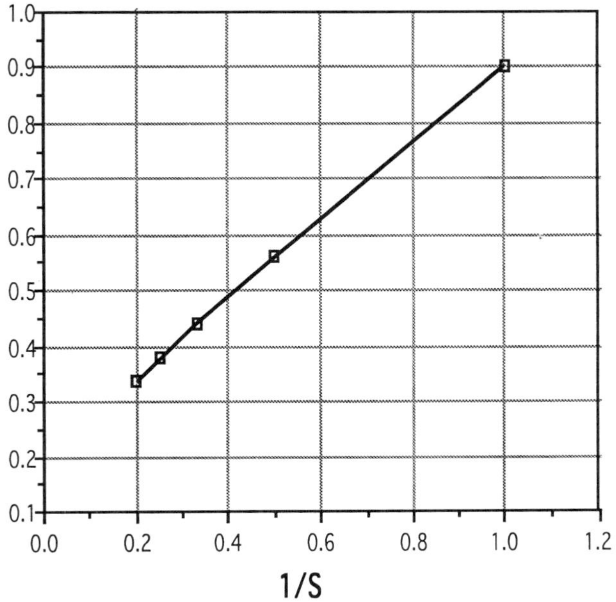

The intercept is 0.19 which is $1/\nu_{max}$ Therefore the maximum velocity of the reaction is **5.0 mmol $L^{-1}s^{-1}$**. The slope is 6.6 x 10^2, which implies that

$$\frac{K_M}{k_b[E_o]} = 6.6 \times 10^2 \text{ mol L}^{-1}\text{s}$$

$$K_M = \frac{6.6 \times 10^2 \text{ mol L}^{-1}\text{ s}}{5.0 \times 10^{-6} \text{ mol L}^{-1}\text{ s}} = \mathbf{1.3 \times 10^{-4}}$$

Maximum velocity = $k_b[Eo]$ = 5.0 mmol $L^{-1}s^{-1}$.
= k_b x (12.5 mmol L^{-1})

k_b = **0.40 s^{-1}**

7.38 To determine if the reaction is competitive or noncompetitive, a plot as in Prob. 7.37 is constructed. If the lines have the same intercept the reaction is competitive. If the lines have different intercepts then the reaction is noncompetitive. It is easy to see from the data that the lines are approximately parallel, and therefore the reaction is **noncompetitive**.

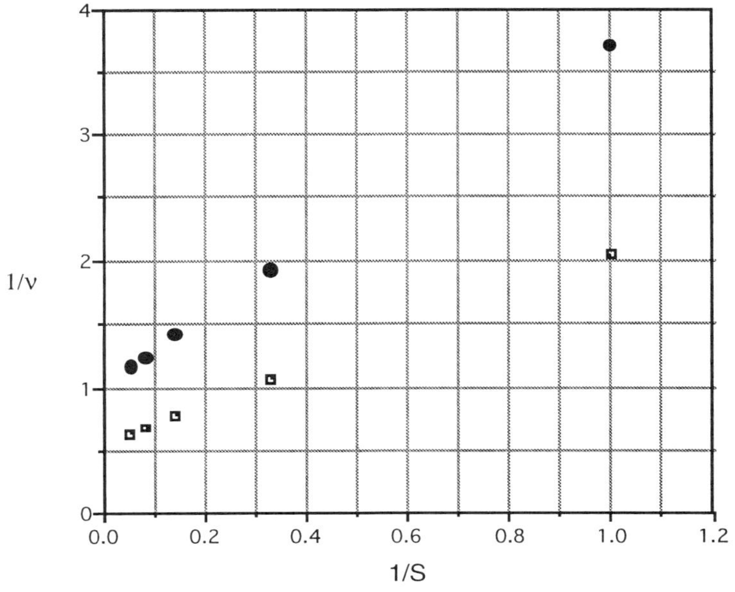

Additional problems

7.1 The rate constant for the reaction

$A_2 + B_2 \rightarrow 2AB$ is 2.7×10^{-4} L mol^{-1} s^{-1} at 600 K and 3.5×10^{-3} L mol^{-1} s^{-1} at 650 K. Find the activation energy, E_a.

7.2 What is the rate constant for the above reaction at 700 K ?

7.3 The reaction
$SO_2Cl_2\ (g) \rightarrow SO_2(g) + Cl_2(g)$
has a rate constant of 9.3×10^{-6} s^{-1}. How long would it take for the concentration of SO_2Cl_2 to decrease 25 percent of its initial value ?

7.4 The reaction

$H_2(g) + I_2(g) \rightarrow 2HI(g)$
may have the following mechanism

$2I \rightleftharpoons 2I$ (fast)

$2I + H_2 \rightarrow 2HI$ (slow)
What rate law is predicted by this mechanism ?

7.5 For the reaction $A \rightarrow B$
the rate is constant and does not change with concentration of A. What can be concluded about the order of the reaction?

7.6 For the above reaction the rate doubles when the initial concentration of A doubles. What is the order of the reaction with respect to A?

7.7 The rate law for the reaction

$A_2 + B \rightarrow C + D$

is rate = $k[A_2]^{1/2}[B]$
What are the units for k?

7.8 For an enzyme reaction, the rate of reaction doubles when the temperature is increased from 25° C to 35° C. What is the activation energy for the reaction/

7.9 A first order reaction is 45.0 percent complete in 65 s. What is the rate constant and the half-life of this reaction?

7.10 What are the units for each of the following if concentrations are expressed in mol L^{-1} ?
(a) rate of a chemical reaction
(b) rate constant for a zero-order rate law
(c) rate constant for a third-order rate law

7.11 For the reaction

$$2NO_2(g) + F_2(g) \rightarrow 2NO_2F(g)$$

The experimentally determined rate law is

rate = $k[NO_2][F_2]$
a suggested mechanism for this reaction is

$NO_2 + F_2 \rightarrow NO_2F + F$ (slow)

$F + NO_2 \rightarrow NO_2F$ (fast)

Is this an acceptable mechanism?

7.12 An inhibitor is added to an enzymatic reaction. As the concentration of inhibitor is increased the slope of a plot of 1/n versus 1/S changes. It the inhibitor competitive or noncompetitive?

Chapter 8

Atomic Structure

8.1 $M = \sigma T^4$

$P = \sigma T^4 \times A = (5.7 \times 10^{-8}\ W\ m^{-2}\ K^{-4}) \times ((3 \times 10^3\ K)^4 \times (10.0 \times 10^{-4}\ m^2)$

$= \underline{\mathbf{4.6 \times 10^3\ W}}$

8.2 $E = P \times t$, $E = nhn$

$$n = \frac{Pt}{h\nu} = \frac{(0.68 \times 10^{-6}\ W) \times (1.0\ s)}{(1.22 \times 10^{15}) \times (6.626 \times 10^{-34}\ J\ Hz^{-1})} = \underline{\mathbf{8.41 \times 10^{11}}}$$

8.3 $p = mv$ and $p = \dfrac{h}{\lambda}$

Therefore,

$$V = \frac{h}{m\lambda} = \frac{(6.626 \times 10^{-34}\ J\ s)}{(9.109 \times 10^{-31}\ kg) \times (0.55 \times 10^{-9}\ m)} = \underline{\mathbf{1.32 \times 10^6\ m\ s^{-1}}}$$

8.4 $p = \dfrac{h}{\lambda}$

(a) $$p = \frac{(6.626 \times 10^{-34}\ J\ s)}{(725 \times 10^{-9}\ m)} = \underline{\mathbf{9.14 \times 10^{-28}\ kg\ m\ s^{-1}}}$$

(b) $$p = \frac{(6.626 \times 10^{-34}\ J\ s)}{(75 \times 10^{-12}\ m)} = \underline{\mathbf{8.8 \times 10^{-24}\ kg\ m\ s^{-1}}}$$

(c) $$p = \frac{(6.626 \times 10^{-34}\ J\ s)}{(20\ m)} = \underline{\mathbf{3.3 \times 10^{-35}\ kg\ m\ s^{-1}}}$$

8.5 $F = DP = \dfrac{(6.626 \times 10^{-34}\ \text{J s})}{(650 \times 10^{-9}\ \text{m})}$ = **1.02×10^{-27} kg m s^{-1}**

b) $p = F/A = (1.02 \times 10^{-27}$ kg m $s^{-1})/(1.0 \times 10^{6}$ $m^{2})$

= **1.02×10^{-33} kg m s^{-1}**

c) $t = \text{speed} \times m/F = \dfrac{1.0\ \text{m s}^{-1} \times 1.0\ \text{kg}}{1.02 \times 10^{-27}\ \text{kg m s}^{-2}}$

= **9.8×10^{26} s**

8.6 $\dfrac{1}{2}mv^2 = hn - I = \dfrac{hc}{\lambda} - I$

$$\lambda = \frac{hc}{I + \frac{1}{2}mv^2}$$

$$= \frac{(6.626 \times 10^{-34}\ \text{J s}) \times (2.998 \times 10^{8}\ \text{m s}^{-1})}{(3.44 \times 10^{-18}\ \text{J}) + \frac{1}{2} \times (9.109 \times 10^{-31}\ \text{kg}) \times (1.03 \times 10^{6}\ \text{m s}^{-1})}$$

$= 5.06 \times 10^{-8}$ m = **50.6 nm**

8.7 $Dp \approx 0.0100$ percent of p_o, and $p_o = m_p v$

$= p_o \times 1.00 \times 10^{-4}$

$$\Delta q \approx \frac{\hbar}{2\Delta p}$$

$$\approx \frac{(1.055 \times 10^{-34}\ \text{J s})}{2 \times (1.673 \times 10^{-27}\ \text{kg}) \times (3.5 \times 10^{5}\ \text{m s}^{-1}) \times (1.00 \times 10^{-4})}$$

$= 9.0 \times 10^{-10}$ m, or **90 nm**

8.8 $E_n = \dfrac{n^2h^2}{8mL^2}$

$$E_2 - E_1 = \frac{(2)^2h^2}{8mL^2} - \frac{(1)^2h^2}{8mL^2}$$

$$= \frac{3 \times (6.626 \times 10^{-34}\ \mathrm{J\,s})^2}{8(1.672 \times 10^{-27}\ \mathrm{kg})(1.0 \times 10^{-9})^2}$$

$= \underline{\mathbf{9.85 \times 10^{-23}\ J}}$

8.9 $\psi = \left(\dfrac{2}{L}\right)^{1/2} \sin\left(\dfrac{n\pi x}{L}\right)$

$$= \left(\frac{2}{L}\right)^{1/2} \sin\left(\frac{\pi x}{L}\right) \quad [n = 1]$$

The wavelength has the greatest value when

$\dfrac{\pi x}{L} = \dfrac{\pi}{2}$, or x = 1/2 L

and at that location

P = 1/2 so the location is 0 -1/2 L or actually, **<u>1/4 L to 3/4 L</u>**

8.10 $\Delta E = (5^2 - 4^2)\dfrac{h^2}{8mL^2} = \dfrac{9h^2}{8mL^2}$

$$= \frac{9 \times (6.626 \times 10^{-34}\ \mathrm{J\,s})^2}{8 \times 9.109 \times 10^{-31}\ \mathrm{kg}\ (5 \times 10^{-9}\ \mathrm{m})^2}$$

$= \underline{\mathbf{2.17 \times 10^{-20}\ J}}$

b) $E = h\nu = \frac{hc}{\lambda}$

l = (6.62608 x 10^{-34} J s) x (2.99792 x 10^{8} m s^{-1})/(2.17 x 10^{-20} J)

= **<u>9.16 x 10^{-6} m</u>**

8.11 $E = h\nu = \frac{hc}{\lambda}$

hc = (6.62608 x 10^{-34} J s) x (2.99792 x 10^{8} m s^{-1}) = 1.986 x 10^{-25} J m

$N_A$hc = (6.02214 x 10^{23} mol^{-1}) x (1.986 x 10^{-25} J m)
= 0.1196 J m mol^{-1}

We can therfore draw up the following table

λ/nm	E/J	E/(kJ mol^{-1})	p/(kg m s^{-1})
(a) 600	3.31 x 10^{-19}	199	1.10 x 10^{-27}
(b) 550	3.61 x 10^{-19}	218	1.20 x 10^{-27}
(c) 400	4.97 x 10^{-19}	299	1.66 x 10^{-27}
(d) 200	9.93 x 10^{-19}	598	3.31 x 10^{-27}
(e) 150 pm	1.32 x 10^{-15}	7.98 x 10^{5}	4.42 x 10^{-24}
(f) 1.00 cm	1.99 x 10^{-23}	0.012	6.63 x 10^{-32}

8.12 $p = \frac{h}{\lambda} = \frac{6.626 \times 10^{-34}\ \text{J s}}{\lambda} = mn$

$$\nu = \frac{(6.626 \times 10^{-34}\ \text{J s})}{(300 \times 10^{-9}\ \text{m}) \times 1.0 \times 10^{-3}\ \text{kg}}$$
= **<u>2.2 x 10^{-24} m s^{-1}</u>**

8.13 $N = \frac{P}{h\nu} = \frac{P\lambda}{hc}$ [P = power in J s^{-1}]

$$= \frac{P\lambda}{(6.626 \times 10^{-34}\ J\,Hz^{-1}) \times (2.998 \times 10^{8}\ m\,s^{-1})}$$

$$= \frac{(P/W) \times (\lambda/nm)\ s^{-1}}{1.99 \times 10^{-16}} = 5.03 \times 10^{15} (P/W) \times (l/nm)\ s^{-1}$$

(a) $N = (5.03 \times 10^{15}) \times 1.0 \times 350\ s^{-1} =$ **$1.7 \times 10^{18}\ s^{-1}$**

(b) $N = (5.03 \times 10^{15}) \times 100 \times 550\ s^{-1} =$ **$1.7 \times 10^{20}\ s^{-1}$**

8.14 $N = \frac{P}{h\nu} = \frac{45 \times 10^{3}\ W}{6.626 \times 10^{-34}\ J\,s \times 98.4 \times 10^{6}\ Hz}$

N = **$6.90 \times 10^{29}\ s^{-1}$**

8.15 From Wein's law,

$T\lambda_{max} = 0.29\ cm\,K$

Therefore,

$T = \frac{0.29\ cm\,K}{480 \times 10^{7}\ cm} =$ **6000 K**

8.16 $\frac{1}{2}mv^{2} = h\nu - \Phi = \frac{hc}{\lambda} - \Phi$

$\Phi = 3.43 \times 10^{-19}\ J$

(a) $\frac{hc}{\lambda} = \frac{(6.626 \times 10^{-34}\ J\,s) \times 2.998 \times 10^{8}\ m\,s^{-1}}{(700 \times 10^{-9}\ m)}$

$= 2.7 \times 10^{-19}\ J < \Phi$, **Therefore no ejection occurs.**

(b) $\frac{hc}{\lambda} = 7.95 \times 10^{-19}$ J

Hence $1/2mv^2 = (7.95 - 3.43) \times 10^{-19}$ J = **4.52×10^{-19} J**

$$v = \left(\frac{2 \times 4.52 \times 10^{-19}\ \text{J}}{9.109 \times 10^{-31}\ \text{kg}}\right)^{1/2} = \mathbf{996\ km\ s^{-1}}$$

8.17 $\Delta E = h\nu = \frac{h}{T}$ [T = period]

(a) $DE = \frac{6.626 \times 10^{-34}\ \text{J s}}{10^{-15}\ \text{s}} = 7 \times 10^{-19}$ J

corresponding to $N_A \times (7 \times 10^{-19}$ J) = **400 kJ mol^{-1}**

(b) DE = **3.5×10^{-20} J , 20 kJ mol^{-1}**

(c) DE = **1.4×10^{-33} J, 8×10^{-13} kJ mol^{-1}**

8.18 $\lambda = \frac{h}{p} = \frac{h}{mv}$

(a) $$\lambda = \frac{(6.626 \times 10^{-34}\ \text{J s})}{(1.00\ \text{m s}^{-1}) \times 1.0 \times 10^{-3}\ \text{kg}} = \mathbf{6.6 \times 10^{-31}\ m}$$

(b) $$\lambda = \frac{(6.626 \times 10^{-34}\ \text{J s})}{(1.0 \times 10^{8}\ \text{m s}^{-1}) \times 1.0 \times 10^{-3}\ \text{kg}} = \mathbf{6.6 \times 10^{-39}\ m}$$

(c) $$\lambda = \frac{6.626 \times 10^{-34}\ \text{J s}}{4.003 \times (1.6605 \times 10^{-27}\ \text{kg}) \times (1.0 \times 10^{3}\ \text{m s}^{-1})} = \mathbf{99.7\ pm}$$

8.19 $\frac{1}{2}mv^2 = e\Delta\phi$, impling that $v = \left(\frac{2e\Delta\phi}{m}\right)^{1/2}$ and

$p = mv = (2me\,\Delta\phi)^{1/2}$, Therefore,

$$\lambda = \frac{h}{p} = \frac{h}{(2me\,\Delta\phi)^{1/2}}$$

$$= \frac{(6.626 \times 10^{-34}\ \text{J s})}{\{2 \times (9.109 \times 10^{-31}\ \text{kg}) \times (1.602 \times 10^{-19}\ \text{C} \times \Delta\phi)\}^{1/2}}$$

$$= \frac{1.226\ \text{nm}}{(\Delta\phi / \text{V})^{1/2}} \qquad [1\ \text{J} = 1\ \text{C V}]$$

(a) $\Delta\phi = 1.00$ V, λ = **<u>1.23 nm</u>**

(b) $\Delta\phi = 1.0$ kV, $\lambda = \dfrac{1.226\ \text{nm}}{31.6}$ = **<u>39 pm</u>**

(c) $\Delta\phi = 100$ kV, $\lambda = \dfrac{1.226\ \text{nm}}{316.2}$ = **<u>3.88 pm</u>**

8.20 $\Delta p \Delta q \geq \frac{1}{2}\hbar, \quad \Delta p = m\Delta v$

$$\Delta V_{min} = \frac{\mathbf{h}}{2m\Delta q} = \frac{1.055 \times 10^{-34}\ \text{J s}}{2 \times 0.500\ \text{kg} \times 5.0 \times 10^{-6}\ \text{m}} = \mathbf{\underline{2.1 \times 10^{-29}\ m\ s^{-1}}}$$

8.21

$$\Delta q_{min} = \frac{\mathbf{h}}{2m\Delta v} = \frac{1.055 \times 10^{-34}\ \text{J s}}{2 \times 0.500\ \text{kg} \times 1.0 \times 10^{-6}\ \text{m s}^{-1}} = \mathbf{\underline{1 \times 10^{-26}\ m}}$$

8.22 $$\Delta p_{min} = \frac{\hbar}{2\Delta v} = \frac{1.055 \times 10^{-34}\ \text{J s}}{2 \times 100 \times 10^{-12}\ \text{m}} = \mathbf{\underline{5 \times 10^{-25}\ kg\ m\ s^{-1}}}$$

$$\Delta v_{min} = \frac{\Delta p_{min}}{m_c} = \frac{5 \times 10^{-25}\ \text{kg m s}^{-1}}{9.109 \times 10^{-31}\ \text{kg}} = \mathbf{\underline{5 \times 10^{5}\ m\ s^{-1}}}$$

8.23 $\frac{1}{2}mv^2 = h\nu - I, \quad \nu = \frac{c}{\lambda}$

$$I = \frac{hc}{\lambda} - \frac{1}{2}mv^2 = \frac{(6.626 \times 10^{-34}\ \text{J s}) \times 2.998 \times 10^8\ \text{m s}^{-1}}{150 \times 10^{-12}\ \text{m}}$$

$$-\frac{1}{2}(9.109 \times 10^{-31}\ \text{kg})(2.24 \times 10^7\ \text{ms}^{-1})^2$$

$$= \mathbf{\underline{1.11 \times 10^{-15}\ J}}$$

8.24 $\rho = \frac{8\pi hc}{\lambda^5}\left(\frac{1}{e^{hc/\lambda kT} - 1}\right) \qquad \Delta U = \rho\Delta\lambda,\ \lambda \approx 652.5\ \text{nm}$

$$\frac{hc}{\lambda k} = \frac{1.439 \times 10^{-2}\ \text{m K}}{\lambda} = 2.205 \times 10^4\ \text{K}$$

$$\frac{8\pi hc}{\lambda^5} = \frac{8\pi \times (6.626 \times 10^{-34}\ \text{J s}) \times 2.998 \times 10^8\ \text{m s}^{-1}}{(652.5 \times 10^{-9}\ \text{m})^5}$$
$$= 4.221 \times 10^7\ \text{J m}^{-4}$$

$$\Delta U = 4.221 \times 10^7\ \text{J m}^{-4} \times \left(\frac{1}{e^{2.205 \times 10^4\ \text{K}/T} - 1}\right) \times (5 \times 10^{-9}\ \text{m})$$

(a) $T = 298\ \text{K},\ \Delta U = \left(\frac{0.211\ \text{J m}^{-3}}{e^{2.205 \times 10^4/298} - 1}\right) = \mathbf{\underline{1.6 \times 10^{-33}\ J\ m^{-3}}}$

(b) $T = 3273\ \text{K},\ \Delta U = \left(\frac{0.211\ \text{J m}^{-3}}{e^{2.205 \times 10^4/3273} - 1}\right) = \mathbf{\underline{2.5 \times 10^{-4}\ J\ m^{-3}}}$

8.25 $\lambda_{max}T = hc/5k$

Therefore

$$\lambda_{max} = \frac{hc}{5k} x \frac{1}{T}$$

if we plot λ_{max} against 1/T we can obtain h from the slope. Draw up the following table:

θ/°C	1000	1500	2000	2500	3000	3500
T/K	1273	1773	2273	2773	3273	3773
10^4/(T/K)	7.86	5.64	4.40	3.61	3.06	2.65
λ_{max}/nm	2181	1600	1240	1035	878	763

The points are plotted in the figure below. The slope is 2.83×10^6, so

$$\frac{hc}{5k} = 2.83 \times 10^6 \text{ nm K} = 2.83 \times 10^{-3} \text{ m K}$$

and

$$h = \frac{5 x (1.38066 x 10^{-23} \text{ J K}^{-1}) x (2.83 x 10^{-3} \text{ m K})}{(2.99792 x 10^8 \text{ m s}^{-1})}$$

$= \underline{\mathbf{6.52 \times 10^{-34} \ J \ s}}$

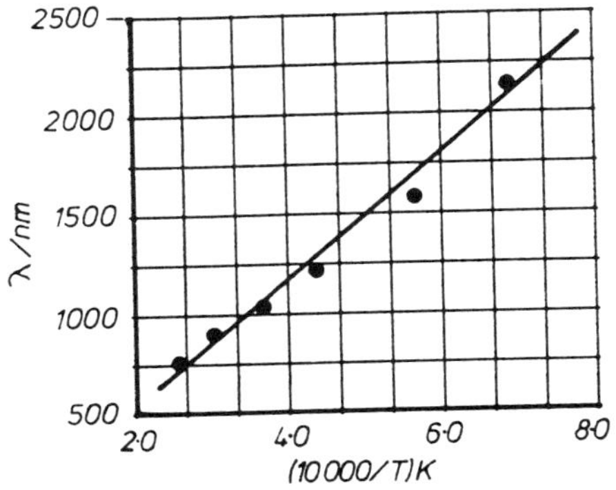

8.26 $\frac{1}{\lambda} = \Re_H \left(\frac{1}{n_1^2} - \frac{1}{n_2^2} \right)$

and $n_1 = 2$, $n_2 = 5$

hence

$$\lambda = (1.09677 \times 10^7\ m^{-1})^{-1}\left(\frac{1}{4}-\frac{1}{25}\right)$$

= **434 nm**

8.27 $\frac{1}{\lambda} = \Re_H\left(\frac{1}{9}-\frac{1}{n^2}\right)$

and therefore

$$n = \left(\frac{1}{9}-\frac{1}{\lambda\Re_H}\right)^{-1/2} = \left(\frac{1}{9}-\frac{\nu}{c\Re_H}\right)^{-1/2}$$

$$= \left(\frac{1}{9}-\frac{2.7415 \times 10^{14}\ s^{-1}}{(2.9979 \times 10^8\ m\ s^{-1}) \times (1.0968 \times 10^7\ m^{-1})}\right)^{-1/2} = 6.005$$

Therefore **n = 6.**

8.28 $\frac{1}{\lambda} = \frac{1}{486.1 \times 10^{-7}\ cm} = 20572\ cm^{-1}$

Hence, the term lies at
T = 27414 cm^{-1} - 20572 cm^{-1} = **6842 cm^{-1}**

b) $E = \frac{hc}{\lambda} = (6.626 \times 10^{-34}\ J\ s) \times 2.998 \times 10^8\ m\ s^{-1} \times 6842\ cm$

E = **1.36 x 10^{-19} J**

8.29 $h\nu = \frac{1}{2}m_e v^2 + I$

$$I = h\nu - \frac{1}{2}m_e v^2 = 6.626 \times 10^{-34}\ \text{J Hz}^{-1} \times \frac{2.998 \times 10^{8}\ \text{m s}^{-1}}{58.4 \times 10^{-9}\ \text{m}}$$

$$-1/2 \times (9.109 \times 10^{-31}\ \text{kg}) \times (1.59 \times 10^{6}\ \text{m s}^{-1})^2$$

$= 2.25 \times 10^{-18}$ J, corresponding to **14.0 eV**

8.30 Since $\psi_{3,0} \propto 6 - 6\rho + \rho^2$, the radial nodes occur at

$6 - 6\rho + \rho^2 = 0$ or $\rho^2 = 3 \pm \sqrt{3}$ = 1.27 and 4.73

Since $\rho = 3\rho a_0 / 2$, the radial nodes occur at **101 pm and 376 pm.**

8.31 nodal planes at **0, 90°, 180° or 270°**
Because $\sin\theta$ goes to zero when θ = 0, 180° and $\cos\theta$ goes to zero when θ = 90, 270°.

8.32 Identify l and use angular momentum = $\{l\ (l + 1)\}^{1/2}\hbar$

(a) l = 0, so **ang. mom. = 0**

(b) l = 0, so **ang. mom. = 0**

(c) l = 2, so **ang mon. = $\sqrt{6}\hbar$**

(d) l = 1, so **ang mon. = $\sqrt{2}\hbar$**

(e) l = 1, so **ang mon. = $\sqrt{2}\hbar$**

The total number of nodes is equal to n - 1 and the number of angular nodes is equal to l ; hence the number of radial nodes is equal to n - 1 - l . We can draw up the following table:

	1s	3s	3d	2p	3p
n, *l*	1,0	3,0	3,2	2,1	3,1
Ang. nodes	0	0	2	1	1
Rad. nodes	0	2	0	0	1

8.33 The energies are $E = -\frac{hc\Re_H}{n^2}$, and the orbital degeneracy g of an energy level of principal quantum number n is $g = n^2$

(a) $E = -hc\Re_H$ implies that n = 1, so **g = 1** (the 1s orbital)

(b) $E = -\frac{hc\Re_H}{9}$ implies that n = 3, so **g = 9** (the 3s orbital, the three p orbitals, and the five 3d orbitals)

(c) $E = \frac{-hc\Re_H}{49}$ implies that n = 7, so **g = 49** (the 7s orbital, the three 7p orbitals, the five 7d orbitals, the seven 7f orbitals, the nine 7g orbitals).

8.34 The probability density varies as

$$\psi^2 = \frac{1}{\pi a_0^3} e^{-2r/a_0}$$

The maximum value is at r = 0 and ψ^2 is 25 per cent of the maximum when

$e^{-2r/a_0} = 0.25$, so that $r = 1/2\, a_o \ln(0.25)$ which is at $\mathbf{r = 0.693\, a_o}$ (37 pm)

8.35 The radial distribution function varies as

$$P = 4\pi r^2 \psi^2 = \frac{4}{a_0^3} e^{-2r/a_0}$$

The maximum value of P occurs at $r = a_o$ since

$$\frac{dP}{dr} \propto \left(2r - \frac{2r^2}{a_0}\right)e^{-2r/a_0} = 0 \text{ at } r = a_o \text{ and } P_{max} = \frac{4}{a_0}e^{-2}$$

P falls to a fraction f of its maximum when

$$f = \frac{\frac{4}{a_0^3}e^{-2r/a_0}}{\frac{4}{a_0}e^{-2}} = \frac{r^2}{a_0^2}e^2e^{-2r/a_0}$$

Therefore solve

$$\frac{f^{1/2}}{e} = \left(\frac{r}{a_0}\right)e^{-r/a_0}$$

(a) $f = 0.25$
solves to $r = 0.7569\ a_o$ or $0.2431\ a_o$ = **<u>40 pm or 13 pm</u>**

(b) $f = 0.10$

solves to $r = 0.554\ a_o$ or $0.446\ a_o$ = **<u>29 or 24 pm</u>**

8.36 Six lobes so probablility is 1/6.

8.37 The selection rules are Dn = any integer; Dl = $\pm$ 1
(a) $2s \rightarrow 1s$; Dl = 0; **<u>forbidden</u>**

(b) $2p \rightarrow 1s$; Dl = -1; **<u>allowed</u>**

(c) $3d \rightarrow 2p$; Dl = -1; **<u>allowed</u>**

(d) $5d \rightarrow 2s$; Dl = -2; **<u>forbidden</u>**

(e) $5p \rightarrow 3s$; Dl = -1; **<u>allowed</u>**

8.38 For a given l there are $2l + 1$ values of m_l and hence $2l + 1$ orbitals. Each orbital may be occupied by two electrons. Therefore the maximum occupancy is $2(2l + 1)$.

	l	$2(2l + 1)$
(a)	0	2
(b)	3	14
(c)	5	22

8.39 Use the building up principle with the orbitals occupied in the order 1s,2s, 2p, 3s, 3p

H	He
$1s^1$	$1s^2$

Li	Be	B	C	N	O
$K2s^1$	$K2s^2$	$K2s^22p^1$	$K2s^22p^2$	$K2s^2sp^3$	$K2s^2sp^4$

F	Ne
$K2s^22p^5$	$K2s^22p^6$

Na	Al	Si	P	S
$KL3s^1$	$KL3s^2$	$KL3s^23p^1$	$KL3s^23p^2$	$KL3s^23p^3$

Mg	Cl	Ar
$KL3s^23p^4$	$KL3s^23p^5$	$KL3s^23p^6$

Where $K = 1s^2$, $L = 2s^2sp^6$

8.40 $$\frac{1}{\lambda} = \Re_H\left(\frac{1}{n_1^2} - \frac{1}{n_2^2}\right) \qquad \Re_H = 109677\ \text{cm}^{-1}$$

Find n_1 from the value of λ_{max}, which arises from the transition $n_1 + 1 \rightarrow n_1$;

$$\frac{1}{\lambda_{max}\Re_H} = \frac{1}{n_1^2} - \frac{1}{(n_1+1)^2} = \frac{2n_1+1}{n_1^2(n_1+1)^2}$$

$$\lambda_{max}\Re_H = \frac{n_1^2(n_1+1)^2}{2n_1+1}$$

$= (12368 \times 10^{-9}\ m) \times (109677 \times 10^2\ m^{-1}) = 135.65$

Since n_1 = 1,2,3 and 4 have already been accounted for try n_1 = 5, 6 ....

With $n_1 = 6$ we get $\frac{2n_1+1}{n_1^2(n_1+1)^2} = 136$. The Humphrey's series is therefore $\mathbf{n_2 \to \underline{6}}$ and the transitions are given by

$$\frac{1}{\lambda} = 109677\ cm^{-1} \times \left(\frac{1}{36} - \frac{1}{n_2^2}\right),\ n_2 = 7,8......$$

and occur at **<u>12370 nm, 7503 nm, 5908 nm, 5129 nm, ... 3908 nm</u>** (at $n_2 = 15$), converging to 3282 as $n_2 \to \infty$.

8.41 $l_{max}(H) = 12370$ nm

$E = hc/l$

$E_{He} \propto Z^2E_H \propto 4E_H$

$l_{He} \propto 1/4\ l_H = (12370\ nm)/4 =$ **<u>3092 nm</u>**

8.42 $\lambda_{max}\Re_H = \frac{n_1^2(n_1+1)^2}{2n_1+1}$ [Problem 8.36]

$= (656.46 \times 10^{-9}\ m) \times (109677 \times 10^2\ m^{-1}) = 7.20$

and hence $n_1 = 2$. Therefore the transitions are given by

$$\frac{1}{\lambda} = 109677\ cm^{-1} \times \left(\frac{1}{4} - \frac{1}{n_2^2}\right)$$

The next line has $n_2 = 7$ and occurs at

$$\frac{1}{\lambda} = 109677\ \text{cm}^{-1} \times \left(\frac{1}{4} - \frac{1}{49}\right) = \underline{\mathbf{397.13\ nm}}$$

The energy required to ionize the atom is obtained by letting $n_2 \to \infty$

Then

$$\frac{1}{\lambda_\infty} = 109677\ \text{cm}^{-1} \times \left(\frac{1}{4} - 0\right) = 27419\ \text{cm}^{-1}, \text{ or } \underline{\mathbf{3.40\ eV}}$$

8.43 $\frac{1}{\lambda} = K \times \left(1 - \frac{1}{n^2}\right)$, n = 2,3, . . .

Therefore if the formula is apropriate we expect to find that $\lambda^{-1}(1 - 1/n^2)^{-1}$ is a constant K. Therefore draw up the following table

n	2	3	4
λ^{-1}	740747	877924	925933
$\lambda^{-1}(1-1/n^2)^{-1}$	987663	987665	987662

Hence the formula does describe the transitions and K = $\underline{\mathbf{987663\ cm^{-1}}}$

Balmer transitions lie at

$$\frac{1}{\lambda} = K \times \left(\frac{1}{4} - \frac{1}{n^2}\right) \qquad n = 3, 4, \ldots.$$

$$= 987663\ \text{cm}^{-1} \left(\frac{1}{4} - \frac{1}{n^2}\right) = \underline{\mathbf{137175\ cm^{-1}}}, \underline{\mathbf{185187\ cm^{-1}}}, \ldots$$

The ionization energy of the ground state ion is given by

$$\frac{1}{\lambda} = K x\left(1 - \frac{1}{n^2}\right),\ n \to \infty$$

and hence corresponds to

$\frac{1}{\lambda}$ = 987663 cm^{-1}, or **122.5 eV**

8.44 a) Periodic Table

Li	Be					B	C	N	O	F	Ne	Na	Mg
Al	Si					P	S	Cl	Ar	K	Ca	Sc	Ti
		Mn	Fe	Co	Ni								
$3s^2$				$3d^4$									

b) Mg and Ti would be Noble gases.

Additional problems

8.1 The red spectral line of lithium occurs at 671 nm. What is the energy of one photon of this light?

8.2 What is the wavelength of radiation emitted when the electron in a hydrogen atom undergoes a transition from $n = 4$ to $n = 2$?

8.3 For potassium, the photoelectric work function equals 3.59×10^{-19} J. What is the minimum frequency of radiation required to eject electrons from potassium?

8.4 What is the wavelength associated with electrons accelerated by 4.00×10^5 V ?

8.5 The peak of a distant star's emission occurs at about 530 nm. Estimate the temperature of its surface.

8.6 Calculate the linear momentum of photons of wavelength (a) 600 nm (b) 10 pm.

8.7 What is the wavelength for an electron ($m = 9.11 \times 10^{-31}$ kg) traveling at a speed of 1.0×10^7 m s^{-1} ?

8.8 The hydrogen atom has a radius on the order of 0.05 nm. Assume the position is known with 1 percent accuracy. Calculate the uncertainty in the velocity.

8.9 Calculate the value of the wavefunction, y^2 for a 1s orbital at $r = 0$ and $r = 2a_o$.

8.10 What is the physical significance of the value of y^2 at a particular point?

Chapter 9

The chemical bond

9.1

(a) Li_2 (6 electrons): $1s\sigma_g^2 1s\sigma_u^2 2s\sigma_g^2$ B.O. = 1

(b) Be_2 (8 electrons): $1s\sigma_g^2 1s\sigma_u^2 2s\sigma_g^2 2s\sigma_u^2$ B. O. = 0

(c) C_2 (12 electrons): $1s\sigma_g^2 1s\sigma_u^2 2s\sigma_g^2 2s\sigma_u^2 2p\pi_g^4$ B. O. =2

9.2 (a) H_2^- (3 electrons) $1s\sigma_g^2 1s\sigma_u^1$ B. O. =0.5

(b) N_2 (14 electrons): $1s\sigma_g^2 1s\sigma_u^2 2s\sigma_g^2 2s\sigma_u^2 2p\pi_g^4 2p\sigma_g^2$
B.O. = 3

(c) O_2 (16 electrons):
$1s\sigma_g^2 1s\sigma_u^2 2s\sigma_g^2 2s\sigma_u^2 2p\sigma_g^2 2p\pi_g^4 2p_{xy}\pi_g^2$
B.O. = 3.

9.3 (a) CO (14 electrons): $1s\sigma^2 1s\sigma^{*2} 2s\sigma^2 2s\sigma^{*2} 2p\sigma^2 2p\pi^4$

(b) NO (15 electrons):
$1s\sigma^2 1s\sigma^{*2} 2s\sigma^2 2s\sigma^{*2} 2p\sigma^2 2p\pi^4 2p\pi^{*1}$

(c) CN^- (14 electrons): $1s\sigma^2 1s\sigma^{*2} 2s\sigma^2 2s\sigma^{*2} 2p\sigma^2 2p\pi^4$

9.4 The bond orders of B_2 and C_2 are respectively 1 and 2. Therefore **C_2** should have the greater bond dissociation enthalpy. The experimental values are approximately 4 eV and 6 eV respectively.

9.5 Decide whether the electron addded or removed increases or decreases the bond order. The simplest procedure is to decide whether the electron occupies or is removed from a bonding or antibonding orbital.

The following table gives the orbital involved

	N_2	NO	O_2	C_2	F_2	CN
(a) AB^-	$2p\pi^*$	$2p\pi^*$	$2p\pi^*$	$2p\sigma$	$2p\sigma^*$	$2p\sigma$
Chng. B.O.	-1/2	-1/2	-1/2	+1/2	-1/2	+1/2
(b)AB^+	$2p\sigma$	$2p\pi^*$	$2p\pi^*$	$2p\pi$	$2p\pi^*$	$2p\sigma$
Chng. B.O.	-1/2	+1/2	+1/2	-1/2	+1/2	-1/2

Therefore **C_2 and CN are stabilized by anion formation. NO, O_2, and F_2 are stabilized by cation formation.**

9.6 For CO we accommodate 14 electrons and for XeF we accommodate 15 valence electrons. The orbitals are sketched below:

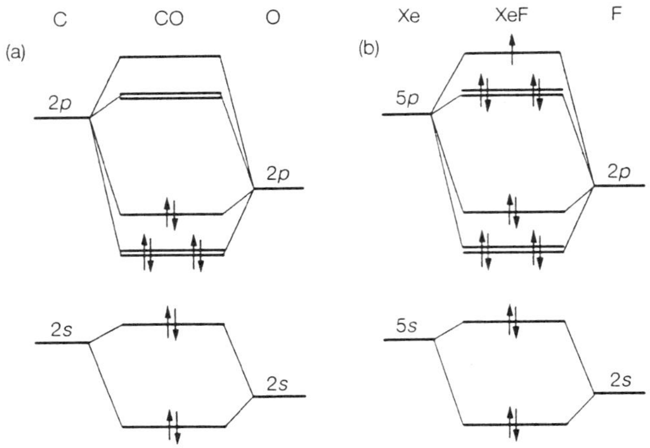

Since the bond order is increased when XeF^+ is formed from XeF (an electron is removed from an antibonding orbital), **XeF^+ will have a shorter bond length than XeF.**

9.7 (a) p* is **g**
(b) g, u is **inapplicable** to a heteronuclear molecule since it has no center of inversion.
(c) **g** (See Fig. below)

(d) **u** (See Fig. below).

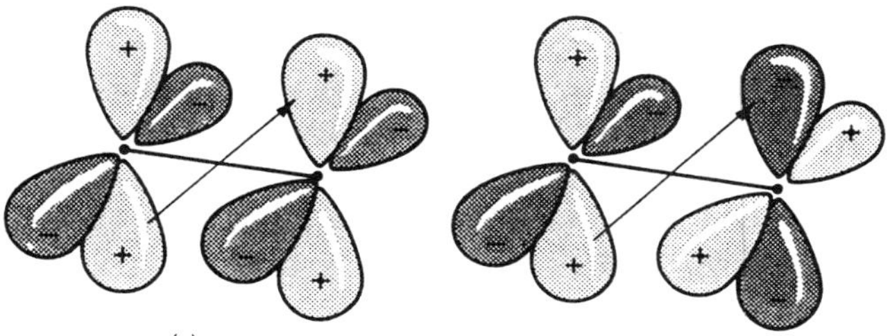

(a) g (g) u

9.8 n = 1 g
n = 2 u
n = 3 g
n = 4 u

9.9 The plane view of the orbitals should be interpreted with the shapes of the p orbitals considered and their nodal planes that lie in the plane of the molecule. The a_2 orbitals are therefore g, the e_1 orbitals are u, and the b_2 orbitals are g.

9.10 The bond orders of NO and N_2 are 2.5 and 3 respectively. Therefore **N_2 should have the shorter bond length.** The experimental values are 115 pm and 110 pm, respectively.

9.11 F_2^- 19 electrons, writing the electron configuration:

$1ss^2 1ss^{*2} 2ss^2 2ss^{*2} 2pp^4 2ps^2 2pp^{*4} 2ps^{*1}$

10 bonding, 9 antibonding

F_2 18 electrons, one less antibonding

F_2^+ 17 electrons, 10 bonding, 7 antibonding

Order of bond lengths: $\mathbf{F_2^+ < F_2 < F_2^-}$

9.12 $$\int \psi^2 d\tau = \frac{1}{3}\int (s+\sqrt{2p})^2 d\tau$$
$$= \frac{1}{3}\int (s^2 + 2p^2 + 2\sqrt{2sp})d\tau$$
$$= \frac{1}{3}(1+2+0) = 1$$
since $\int s^2 d\tau = 1$, $\int p^2 d\tau = 1$, and $\int sp\, d\tau = 0$ [orthogonality]

9.13 See Prob. 9.12.

$$\int \frac{(s+\sqrt{2p})}{\sqrt{3}} \Psi_B d\tau = 0$$

$$\Psi_B = \frac{(\sqrt{2}s - \sqrt{2p})}{\sqrt{3}}$$ Will be orthogonal.

9.14 $$\int \psi^2 d\tau = N^2 \int (\psi_A + \lambda\psi_B)^2 d\tau$$
$$= N^2 \int (\psi_A^2 + \lambda^2\psi_B^2 + 2\lambda\psi_A\psi_B) d\tau$$
$$= N^2(1+\lambda^2 - 2\lambda S) \quad \left[\int (\psi_A\psi_B) d\tau = S\right]$$
Therefore we require
$$N = \left(\frac{1}{1+2\lambda S+\lambda^2}\right)^{1/2}$$

9.15 (a) CO_2 is **linear**, either by VSEPR theory (two atoms attached to the central atom, no lone pairs on C) or by regarding the molecule as having a s framework and p bonds between the C and O atoms.

(b) NO_2 is **angular** since it is isoelectronic with CO_2^-. The extra electron is a "half lone pair" and a bending agent. The extra electron can be accommodated by the molecule bending so as to give the lone pair some s orbital character.

(c) NO_2^+ is **linear,** since it is isoelectronic with CO_2.

(d) NO_2^- is **angular** since it has one more electron than NO_2 and a corresondingly stronger bonding influence.

(e) SO_2 is **angular** it is isoelectronic with NO_2^-.

(f) H_2O is **angular** The extra electron pair will act as a bending agent.

(g) H_2O^{2+} is **angular** since one electron pair responsible for the bending is still present.

9.16 (a) H_2S is **angular**. 6 valence electrons for sulfur lead to two lone pairs. The electron arrangement is tetrahedral, the molecular shape is angular.
(b) SF_6 Six electron pairs give **octahedral** shape.

(c) XeF_4 Six electron pairs give octahedral shape. Two lone pairs at 180° lead to **square planar** shape for the molecule.

(d) 10 outer shell electrons and five electron pairs arranged as trigonal bipyramid. Four pairs are bonding and one is a lone pair. This gives a distorted **see-saw** shape (text p.375).

9.17 The molecular orbitals of the fragments and the molecular orbitals that they form are shown in the following figure:

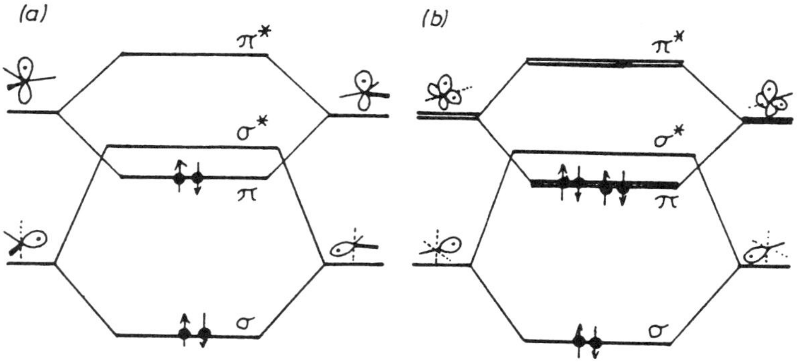

9.18 (a) C_6H_6 (6 electrons): $a_{2u}^2e_{1g}^4$

$E = 2(\alpha+2\beta)+4(\alpha+\beta) = \underline{\mathbf{6}\alpha+\mathbf{8}\beta}$

(b) $C_6H_6^+$ (5 electrons): $a_{2u}^2e_{1g}^3$

$E = 2(\alpha+2\beta)+3(\alpha+\beta) = \underline{\mathbf{5}\alpha+\mathbf{7}\beta}$

9.19 $E_n = \dfrac{n^2h^2}{8mL^2}$, n = 1, 2, . . . and $\psi_n = \left(\dfrac{2}{L}\right)^{1/2}\sin\left(\dfrac{n\pi x}{L}\right)$

Two electrons occupy each level (by the Pauli principle). and so butadiene has two electrons in ψ_1 and two electrons in ψ_2:

$$\psi_1 = \left(\frac{2}{L}\right)^{1/2}\sin\left(\frac{\pi x}{L}\right) \qquad \psi_2 = \left(\frac{2}{L}\right)^{1/2}\sin\left(\frac{2\pi x}{L}\right)$$

The minimum energy is

$$\Delta E = E_3 - E_2 = 5\left(\frac{h^2}{8m_cL^2}\right)$$

Tetraene has eight p electrons to accommodate, so the HOMO

and LUMO will be ψ_4 and ψ_5, respectively. From the particle-in-a-box solutions,

$$\Delta E = E_4 - E_5 = (25-16)\left(\frac{h^2}{8m_cL^2}\right) = \frac{9h^2}{8m_cL^2}$$

$$= \frac{9\text{x}(6.626\text{x}10^{-34}\ \text{J s})^2}{8\text{x}(9.109\text{x}10^{-31}\ \text{kg})\text{x}(1.12\text{x}10^{-9}\ \text{m})^2} = \underline{\mathbf{4.3\ x\ 10^{-19}\ J}}$$

It follows that

$$\lambda = \frac{hc}{\Delta E} = \frac{(6.626\text{x}10^{-34}\ \text{J s})\text{x}(2.998\text{x}10^{8}\ \text{m s}^{-1})}{4.3\text{x}10^{-19}\ \text{J}}$$

$= 4.6 \text{ x } 10^{-7}$ m or **460 nm**

The HOMO and LUMO are $\psi_n = \left(\frac{2}{L}\right)^{1/2} \sin\left(\frac{n\pi x}{L}\right)$ with n = 4,5 respectively and the two wavefunctions are sketched below:

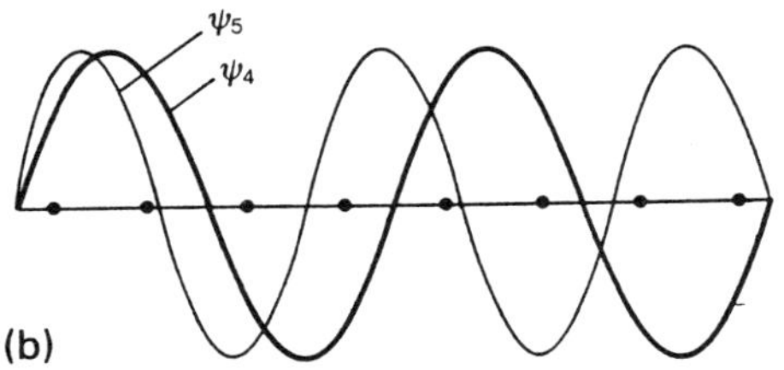

(b)

Additional problems

9.1 Arrange the following bonds in order of increasing polarity: P-H, H-O, and C-Cl.

9.2 For the molecules N_2H_4, N_2 and N_2F_2, which molecule has the shortest nitrogen-nitrogen bond?

9.3 What is the geometry of the ClF_3 molecule?

9.4 Using VSEPR theory, predict which of the following compounds will have the smallest H-X-H bond angle (where X is the central atom): CH_4, NH_3, H_2O.

9.5 Use the molecular orbital model to predict the bond order and stability of Ne_2 and P_2.

9.6 Give the ground state electron configurations and bond order for O_2^+ and O_2^-.

9.7 For the following molecules predict the molecular structure and give the hybrid orbitals for the central atom. a) OF_2 b) CF_4 c) KrF_2

9.8 A Lewis structure obeying the octet rule can be drawn for O_2 as follows:

:O=O:

Use molecular orbital theory to explain why the above Lewis structure corresponds to an excited state.

9.9 Describe the hybrid orbitals used by the carbon atoms in C_3H_5OH.

9.10 The molecule H_2S has a net dipole moment. Is the molecule linear?

Chapter 10

Cohesion and Structure

10.1 In order to estimate the lattice enthalpy, the following series of steps can be written:

$MgO(s) \rightarrow Mg(s) + 1/2O_2(g)$ $\Delta H_f^\ominus$ = +601.70 kJ mol^{-1}

$Mg(s) \rightarrow Mg(g)$ 147.70

$Mg(g) \rightarrow Mg^+(g) + e^-$ 736.0

$Mg^+ \rightarrow Mg^{2+} + e^-$ 1450.0

$1/2O_2(g) \rightarrow O(g)$ 249.0

$O(g) + e^- \rightarrow O^-(g)$ -141.0

$O^-(g) + e^- \rightarrow O^{2-}(g)$ 844.0

Overall:
$MgO(s) \rightarrow Mg^{2+}(g) + O^{2-}(g)$
DH = 3887 kJ mol^{-1} = **<u>3.9 x 10^3 kJ mol^{-1}</u>**

10.2 The ratio of the two expressions for the lattice enthalpy, assuming A is the same in both cases and that (1 - d*/d) is approximately the same for both molecules will give

$$\frac{\Delta H(CaO)}{\Delta H(SrO)} \propto \frac{r_{SrO}}{r_{CaO}} = \frac{256}{240} = \underline{\mathbf{1.07}}$$

10.3 A D_{3h} (trigonal planar) molecule is nonpolar; therefore the more likely structure is **<u>two axial and one equatorial F atoms.</u>**

10.4 Add the dipole moments vectorially
(a) p-xylene: the resultant vector is zero so **$\mu = 0$**

(b) o-xylene: $\mu = 0.4\ \text{D}\cos 30^\circ + 0.4\ \text{D}\cos 30^\circ =$ **0.7 D**

(c) m-xylene: $\mu = 0.4\ \text{D}\cos 60^\circ + 0.4\ \text{D}\cos 60^\circ =$ **0.4 D**

Cancellation of μ's is exact by symmetry so the p-xylene molecule is nonpolar.

10.5 Add the dipole moments vectorially using the dipoles from the preceding problem
(a) 1,2,3-trimethylbenzene: m-xylene + 0.40D = **0.8 D**

(b) 1,2,4-trimethylbenzene: p-xylene + $0.4\ \text{D}\cos 60^\circ$ = **0.2 D**

(c) 1,3,5-trimethylbenzene: m-xylene - 0.4 D = 0

(c) should cancel exactly.

10.6 $\mu = (\mu_1^2 + \mu_2^2 + 2\mu_1\mu_2\cos\theta)^{1/2}$

$= \{(1.5)^2 + (0.80)^2 + 2 \times 1.5 \times 0.80 \times (\cos 109.5^\circ)\}^{1/2}\ \text{D}$

= **1.4 D**

10.7 The three conformations will have dipole moments of 3.0 D, the vectors add; D = 0, the vectors cancel, and $1.5\cos 30^\circ + 1.5\cos 30^\circ = 2.6$ D.

a) **1.87 D**

b) **3.0 D**

c) (2 x 3.0D + 2.6 D)/4 = **2.15 D**

d) (3.0 + 2 x 2.6 D)/5 = **1.64 D**

10.8 The total potential energy of interaction between the dipole μ = q'l and the point charge, q is the sum of repulsion and attraction terms

$$4\pi\varepsilon_o V = \frac{-q'q}{r-\frac{1}{2}l} + \frac{q'q}{r+\frac{1}{2}l}$$

10.9 Use the above expression and
Since l << r

$$4\pi\varepsilon_o V = \frac{-q'q}{r}\left(\frac{1}{1-x} - \frac{1}{1+x}\right)$$

where x = l/2r

$$= \frac{-q'q}{r}\{(1+x+x^2...)-(1-x+x^2....)\}$$

$$= \frac{-q'q}{r}x(2X) = \frac{-q'ql}{r^2} = \frac{-q'\mu}{r^2}$$

$$V = \frac{-q'\mu}{4\pi\varepsilon_o r^2}$$

10.10 Proceed in the same way as Prob. 10.9, but not the total interaction energy is the sum of four pairwise terms

$$4\pi\varepsilon_o V = \frac{q_1q_2}{r+l} - \frac{q_1q_2}{r} - \frac{q_1q_2}{r} + \frac{q_1q_2}{r-l}$$ use x = l/r

$$4\pi\varepsilon_o V = \frac{q_1q_2}{r}\left(\frac{1}{1+x} - 2 + \frac{1}{1-x}\right)$$

Expand the terms in x retaining only the first surviving term

$$\frac{q_1q_2}{r}\{(1+x+x^2...)-2+(1-x+x^2....)\}$$

$$4\pi\varepsilon_oV = \frac{q_1q_2}{r}(2x^2)$$

$$= \frac{2(q_1l)(q_2l)}{r^3} = \frac{2\mu_1\mu_2}{r^3}$$

$$V = \frac{2\mu_1\mu_2}{4\pi\varepsilon_o r^3}$$

10.11 The Lennard-Jones potential can be expressed

$$V = 4\varepsilon\left\{\left(\frac{\sigma}{r}\right)^{12} - \left(\frac{\sigma}{r}\right)^{6}\right\}$$

The minimum occurs where

$$\frac{dV}{dr} = 4\varepsilon\left\{\left(\frac{12\sigma^{12}}{r^{13}}\right) - \left(\frac{6\sigma^6}{r^7}\right)\right\} = 0$$

$$\frac{12\sigma^{12}}{r^{13}} = \frac{6\sigma^6}{r^7}$$

$$r = 2^{1/6}\sigma$$

10.12 In the dimer, the dipole moments will partially cancel, depending on the exact orientation. As the temperature increases, there will be more monomers and the dipole moments will not cancel.

10.13 The potential energy has the form

$$V = 4\varepsilon\left\{Ae^{-r/\sigma} - \left(\frac{\sigma}{r}\right)^6\right\}$$ and is sketched in the figure below:

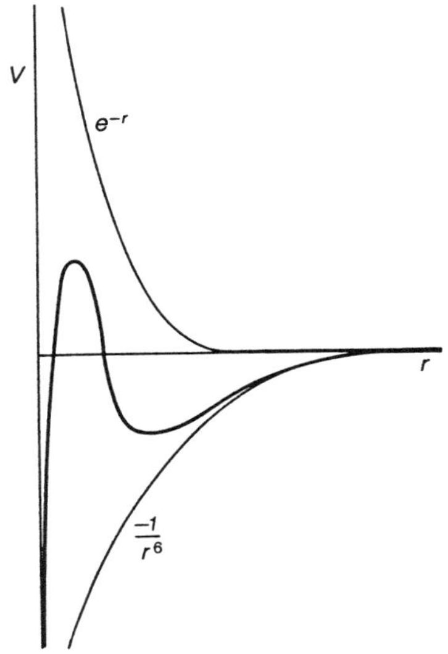

$$\frac{dV}{dr} = 4\varepsilon\left\{-\frac{A}{\sigma}e^{-r/\sigma} + \left(\frac{6\sigma^6}{r^7}\right)\right\} = 0$$

which occurs at the solution of

$$\frac{A}{6}e^{-r/\sigma} = \frac{\sigma^7}{r^7}$$

Solve this equation numerically. As an example, when A = s = 1, a minimum occurs at **r = 1.63**.

10.14 For the virial equation

$$\frac{pV_m}{RT} = 1 + \frac{B}{V_m} + \frac{C}{V_m^2} + \ldots.$$

$C = B^2$

$$\frac{pV_m}{RT} = 1 + \frac{B}{V_m} + \frac{B^2}{V_m^2} + \ldots.$$

This is the expansion for 1/(1-x) where $x = B/V_m$

$$\frac{pV_m}{RT} = \frac{1}{1 - \frac{B}{V_m}}$$

Solving for V_m

$$V_m = \frac{RT}{p} + B$$

10.15 In both cases we view the liquid as hard spheres, then the radial distribution function will resemble that in Fig. 10.13 of the text. For a cubic close-packed arrangement, the maxima will occur at $\frac{a}{\sqrt{2}}$ and a, where a is the length of the unit cell (4r).

The first minimum occurs at $\frac{a}{2\sqrt{2}}$.

For a body-centered cubic, the maxima occur at $\frac{\sqrt{3}}{2}a$ and a. The minima will occur halfway between.

10.16 (a) $D = \frac{\lambda^2}{2\tau} = \frac{(1.5 \times 10^{-10}\ \text{m})^2}{2 \times 1.8 \times 10^{-12}\ \text{s}} =$ $\underline{\mathbf{6.2 \times 10^{-9}\ m^2\ s^{-1}}}$

(b) $D = \frac{(0.75 \times 10^{-10}\ \text{m})^2}{2 \times 1.8 \times 10^{-12}\ \text{s}} =$ $\underline{\mathbf{1.6 \times 10^{-10}\ m^2\ s^{-1}}}$

10.17 $d^2 = 2Dt$

$$D = \frac{\lambda^2}{2\tau}$$ substituting into the first equation gives

$$d^2 = \frac{\lambda^2}{\tau}t \qquad d = \lambda\sqrt{t/\tau}$$

$d = 10^3\lambda$, the number of steps is t/t therefore the number of steps is $d^2 = \underline{\mathbf{10^6}}$

b) number of steps will be the same.

10.18 a) For a stack of cylinders

$f = \frac{NV_a}{V_c}$, $V_a = \pi R^2 L$ $V_c = (2R)^2 \sin 60^\circ$ Since the cylinders will form a hexagonal arrangement.

$$f = \frac{\pi R^2 L}{(2R)^2 \sin 60^\circ} = \underline{\mathbf{0.9069}}$$

b) $N = 1$, $V_a = \frac{4\pi R^3}{3}$, $V_c = (2R)^3$

$$f = \frac{4\pi R^3}{3 x (2R)^3} = \frac{\pi}{6} = \underline{\mathbf{0.5236}}$$

10.19 a) **eight nearest neighbors**

b) **six next-nearest neighbors**

Distances: nearest neighbors--calculate the diagonal of the face of the cube. Then calculate the distance of the right triangle formed by the diagonal (c), one side of the cube (a) and a diagonal (d) from the corner of the base to diagonally opposite corner at the top of the cube.

$a^2 + a^2 = c^2$ and $c^2 + a^2 = d^2$ a= 500 nm, d = 866 nm so the distance is 1/2 d = **433 nm**

The next nearest neighbors will be one length of the cube away, **500 nm**.

10.20 a) **12 nearest neighbors**
b) **20 next nearest neighbors**

nearest neighbors $2a^2 = d^2$ 1/2 d = **354 nm**

next nearest neighbors

$a^2 + (1/2\ d)^2 = (d')^2$ d' = **613 nm**

10.21 a) cubic close packed -> body centered cubic, **less dense** (4 atoms or ions/ unit cell -> atoms or ions/unit cell)

b) density will change by **factor of two.**

10.22 The points and planes are shown in the Fig. below.

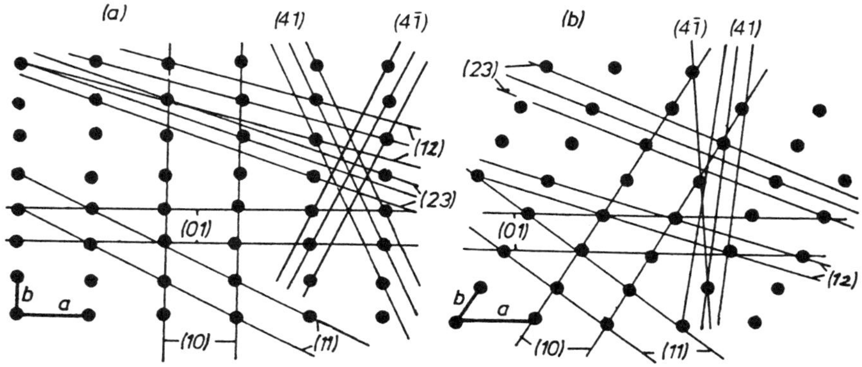

10.23 See Fig. b above.

10.2 4 Draw up the following table

Original	Reciprocal	Clear Fract.	Miller Indices
(2a, 3b, c) or (2,3,1)	(1/2, 1/3, 1)	(3, 2, 6)	(326)
(a, b, c) or (1, 1, 1)	(1, 1, 1)	(1, 1, 1)	(111)
(6a, 3b, 3c) or (6,3,3)	(1/6,1/3,1/3)	(1, 2, 2)	(122)
(2a, -3b, -3c) or (2,-3,-3)	(1/2,-1/3,-1/3)	(3,-2,-2)	(322)

10.25 The planes are drawn in the figure below:

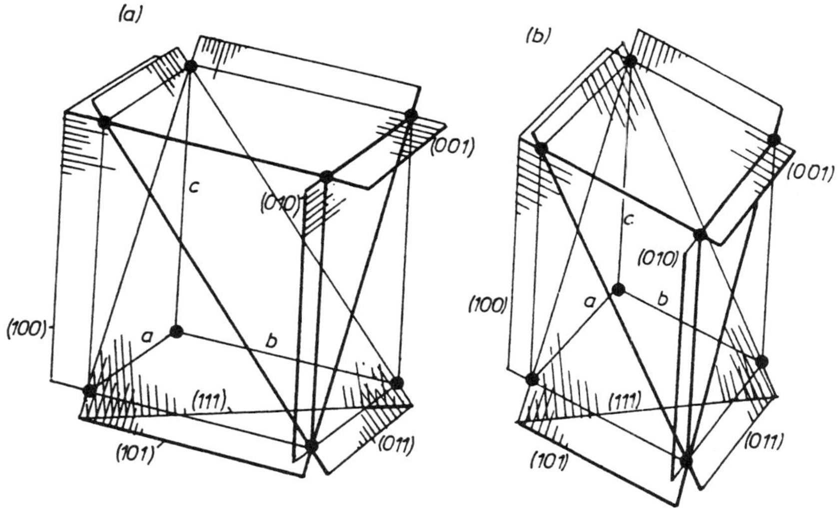

10.26 See the figure above.

10.27 $d_{khl} = \dfrac{a}{(h^2 + k^2 + l^2)^{1/2}}$

Therefore,

$$d_{111} = \frac{a}{3^{1/2}} = \frac{532\text{ pm}}{3^{1/2}} = \underline{\mathbf{307\text{ pm}}}$$

$$d_{211} = \frac{a}{6^{1/2}} = \frac{532\ \text{pm}}{6^{1/2}} = \underline{\mathbf{217\ pm}}$$

$d_{100} = a =$ **<u>532 pm</u>**

10.28 $\lambda = 2d\sin\theta$

$= 2 \times (97.3\ \text{pm}) \times (\sin 19.85^{\circ}) =$ **<u>66.1 pm</u>**

10.29 $\theta = \arcsin\dfrac{\lambda}{2d}$

$$\Delta\theta = \arcsin\frac{\lambda_1}{2d} - \arcsin\frac{\lambda_2}{2d}$$

$$= \arcsin\left(\frac{154.051\ \text{pm}}{2 \times 77.8\ \text{pm}}\right) - \arcsin\left(\frac{154.433\ \text{pm}}{2 \times 77.8\ \text{pm}}\right) = -1.07^{\circ}$$
$= 0.0187$ rad

The separation of the components is therefore 2 x 5.74 cm x 0.0187 = **<u>0.21 cm.</u>**

10.30 V = 651 pm x 651 pm x 934 pm = **<u>3.96 x 10^{-28} m^3</u>**

Each unit cell contains one molecule.

571.81 g mol^{-1} x (1 mol/ 6.023 x 10^{23} atoms)

= 94.94 x 10^{-23} g atom^{-1}

D = m/V = 94.94 x 10^{-23} g /3.96 x 10^{-28} m^3

= **<u>2.40 x 10^6 g m^{-3}</u>**

10.31 $\rho = \dfrac{NM}{VN_A}$ [N is the number of formula units per unit cell]

$$N = \frac{\rho V N_A}{M} =$$

$$\frac{(3.9 \times 10^{6}\ \text{g m}^{-3}) \times (634 \times 784 \times 516 \times 10^{-36}\ \text{m}^{3}) \times (6.022 \times 10^{23}\ \text{mol}^{-1})}{154.77\ \text{g mol}^{-1}}$$

$$= 3.9$$

Therefore, **N = 4** and the true calculated density (in the absence of defects) is

$$\rho = \frac{4 \times (154.77\ \text{g mol}^{-1})}{(634 \times 784 \times 516 \times 10^{-36}\ \text{m}^{3}) \times (6.022 \times 10^{23}\ \text{mol}^{-1})}$$

= **$4.01\ \text{g cm}^{-3}$**

10.32 $d_{hkl}^{-1} = \left\{ \left(\dfrac{h}{a}\right)^2 + \left(\dfrac{k}{b}\right)^2 + \left(\dfrac{l}{c}\right)^2 \right\}^{1/2}$

$$= \left\{ \left(\frac{3}{812}\right)^2 + \left(\frac{2}{947}\right)^2 + \left(\frac{1}{637}\right)^2 \right\}^{1/2}\ \text{pm}^{-1}$$

d = **220 pm**

b) $d_{hkl}^{-1} = \left\{ \left(\dfrac{6}{812}\right)^2 + \left(\dfrac{4}{947}\right)^2 + \left(\dfrac{2}{637}\right)^2 \right\}^{1/2}$

d = **110 pm**

10.33 $d_{100} = a = 350$ pm

$$\rho = \frac{NM}{VN_A} \text{ implying that}$$

$$N = \frac{\rho VN_A}{M} = \frac{(0.56 \times 10^6\ g\,m^{-3}) \times (350 \times 10^{-12}\ m)^3 \times (6.022 \times 10^{23}\ mol^{-1})}{6.94\ g\,mol^{-1}}$$

$$= 1.97$$

An fcc cubic cell has N = 4 and a bcc unit cell has N = 2. Therefore, lithium has a **bcc unit cell.**

10.34 $\theta_{khl} = \arcsin\left\{\frac{\lambda}{2a}\left(h^2 + k^2 + l^2\right)^{1/2}\right\}$

The systematic absences in an fcc structure are that (hkl) all even or all odd are the only permitted lines. Since l/2a = 0.213 we expect the following lines:

(hkl):	111	200	220	311
θ	21°	25°	37°	45°

The density is calculated from

$$\rho = \frac{NM}{VN_A} = \frac{4 \times (63.55\ g\,mol^{-1})}{(361\,pm)^3 \times (6.022 \times 10^{23}\ mol^{-1})} = \mathbf{8.97\ g\ cm^{-3}}$$

Additional problems

10.1 X-rays of wavelength 154 pm were used to analyze an aluminum crystal. A reflection occurred at $\theta = 19.3^\circ$. Calculate the distance between planes (assume n = 1).

10.2 A solid with atoms in a face-centered cubic unit cell has an edge length of 392 pm and a density of 21.45 g/cm^3. What is the molar mass of the solid?

10.3 The density of tungston is 19.3 g/cm^3 and the atomic radius is 139 pm. Does tungston have a body centered or face-centered cubic unit cell?

10.4 Copper metal has a face-centered cubic structure and a density of 8.93 g/cm^3. The molar mass is 63.5 g mol^{-1}. What is the radius of the copper atom?

10.5 Estimate the lattice enthalpy of sodium chloride.

10.6 The molecule CO_2 has no net dipole. What can be concluded about the shape of the molecule?

10.7 Estimate the diffusion constant for a molecule that leaps 10^3 pm each 2.0 ps.

10.8 The following data were obtained for a protein moving through water under the influence of an electric field:

pH	5.0	5.5	6.0	6.5
speed/μm s^{-1}	-3.0	-1.0	+1.0	+3.0

What is the isoelectric point of the protein?

10.9 The self-diffusion coefficient of H_2O in water is 2.26 x 10^{-5} cm^2 s^{-1} at 25° C. How long does it take for an H_2O molecule to travel 10 cm?

10.10 A molecule with the formula $C_2H_2Cl_2$ is nonpolar. What is the structure of the molecule?

Chapter 11

Molecular spectroscopy

11.1 DE = 2hcB(J + 1)

$\overline{\nu}$ = 2B

0.2 J/h = (J + 1)

(0.2 J)/(6.624 x 10^{-34} J) = **$\underline{3.0 \times 10}^{32}$**

11.2 $DE = \dfrac{hc}{\lambda} = \dfrac{hc\mathbf{h}}{4\pi m_H R^2}$

$$\frac{1}{\lambda} = \frac{1.054 \times 10^{-34}\ \text{Js}}{4\pi\left(1.762 \times 10^{-27}\ \text{kg}\right) \times \left(160 \times 10^{-12}\ \text{m}\right)^2} = \underline{\mathbf{5.1\ pm}}$$

11.3 Those with a dipole moment: **<u>a,b, c, and d. e will not.</u>**

11.4 All of them because XeF_4 will have an induced dipole.

11.5 $I_{||} = I_{\perp}$ and A = B

$E_{J.K}$ = hcBJ(J + 1) + hc(A - B)K^2

= hcBJ(J + 1) for methane

Only **<u>one state with J - 10.</u>**

11.6 From the above equation and K = J, J -1, ... -J, There will be **<u>10 rotational states.</u>**

11.7 moment of inertia for octahedral molecule

$I = \sum_J m_J x_J^2$ where we assume the m are equal. If one views the octahedral as a central atom with four addditional (B) atoms in the plane, and one atom above and one below, **$\underline{I = 4m_B r^2}$**

11.8 Square planar molecule would be sum of two linear rotors, $I = 2\ mR^2$ so for square planar, **$\underline{I = 4m_B r^2}$**

11.9 (a) $DE = E_{J+1} - E_J$

$$= hcBJ(J + 1) - hcB(J - 1)(J - 1 + 1)$$

$$= hc2BJ$$

$$= 2hcB$$

$$\frac{1}{\lambda} = \frac{E}{hc} = 2B = \mathbf{\underline{21.2\ cm^{-1}}}$$

b) $\frac{1}{\lambda} = \frac{\nu}{c}$, $\nu = c \times 21.2\ cm^{-1} = \mathbf{\underline{63.6 \times 10^{10}\ Hz}}$

11.10 $DE = 6\ hcB$

$$\frac{1}{\lambda} = \frac{E}{hc} = 6B = \mathbf{\underline{63.6\ cm^{-1}}}$$

$c \times 63.6\ cm^{-1} = \mathbf{\underline{1.91 \times 10^{12}\ Hz}}$

11.11 I would be lareger, therefore B smaller and therefore **<u>lower wavenumber.</u>**

11.12 $B = 0.3904\ cm^{-1}$

$$I = \frac{h}{4\pi cB} = \frac{1.054 \times 10^{-34}\ Js}{4\pi(2.998 \times 10^{8}\ m\ s^{-1}) x (39.04\ m^{-1})}$$

$$= 7.170 \times 10^{-46}\ kg\ m^2$$

$$I = \frac{m_1 m_2}{m_1 + m_2} R^2 = \frac{(1.992 \times 10^{-26}\ kg)(2.6556 \times 10^{-26}\ kg)}{1.992 \times 10^{-26}\ kg + 2.6556 \times 10^{-26}\ kg} R^2$$

R = **<u>251 pm</u>**

11.13 wavenumber = 2B, $B = 6.4\ cm^{-1}$

Solving as above using appropriate masses,

R = **<u>163 pm</u>**

b) If $^2HI^{127}$ calculate new center of mass = 3.27×10^{-27} kg

Assume R is the same,

$$I = \frac{m_1 m_2}{m_1 + m_2} R^2 = 8.68 \times 10^{-47}$$

$$B = \frac{h}{4\pi cI} = 3.22\ cm^{-1}$$

$\bar{\nu} = 2B =$ **<u>$6.44\ cm^{-1}$</u>**

11.14 Select those molecules in which a vibration gives rise to a change in dipole moment. They are **(b) HCl, (c) CO_2, (d) H_2O, (e) CH_3CH_3 (f) CH_4 (g) CH_3Cl.**

11.15 $\omega = \left(\dfrac{k}{\mu}\right)^{1/2}$ and $\omega = 2\pi\nu = 2\pi c\tilde{\nu}$

Therefore $k = \mu\omega = 4\pi^2 c^2 \tilde{\nu}^2, \quad \mu = \frac{1}{2}m(^{35}Cl)$

$$k = 4\pi^2 \frac{34.9688}{2} x (1.66054 x 10^{-27}\text{ kg})$$

$$x\{(2.99792 x 10^{10}\text{ cm s}^{-1}) x (564.9\text{ cm}^{-1})\}^2$$

= **$\underline{328.7\text{ N m}^{-1}}$**

11.16 $\lambda_{obs} = \left(1 + \dfrac{v}{c}\right)\lambda$ When using this formula, take $v > 0$ for recession and $v < 0$ for approach. 50 m. p.h. corresponds to 24.6 m s^{-1}.

$$\lambda_{obs} = \left(1 - \frac{24.6\text{ m s}^{-1}}{2.998 x 10^8\text{ m s}^{-1}}\right) x 660\text{ nm}$$

= **$\underline{0.999\ 999\ 18 \text{ x } 660\text{ nm}}$**

$$v = \left(\frac{\lambda_{obs}}{\lambda} - 1\right)c = (2.998 x 10^8\text{ m s}^{-1}) x \left(\frac{520\text{ nm}}{660\text{ nm}} - 1\right)$$

= **$\underline{6.36 \text{ x } 10^7\text{ m s}^{-1}}$** or about 1.4 x 10^8 m. p. h.

11.17 $v = \left(\dfrac{\lambda_{obs}}{\lambda} - 1\right)c$

$$= (2.998 x 10^8\text{ m s}^{-1}) x \left(\frac{706.5\text{ nm}}{654.2\text{ nm}} - 1\right) = \mathbf{\underline{2.4 \text{ x } 10^4\text{ km s}^{-1}}}$$

$$\delta\lambda = \frac{2\lambda}{c}\left(\frac{2kT}{m}\ln 2\right)^{1/2}$$ which implies that

$$T = \frac{m}{2k\ln 2}\left(\frac{c\delta\lambda}{2\lambda}\right)^2$$

$$= \frac{48\times(1.6605\times10^{-27}\ \text{kg})}{2\times(1.381\times10^{-23}\ \text{J K}^{-1})\times\ln 2}\left(\frac{(2.998\times10^{8}\ \text{m s}^{-1})\times(61.8\times10^{-12}\ \text{m})}{2\times(654.2\times10^{-9}\ \text{m})}\right)^2$$

= **<u>8.4 x 10⁵ K</u>**

11.18 $\delta\tilde{\nu} = \dfrac{5.3\ \text{cm}^{-1}}{\tau/\text{ps}}$ implying that $\tau = \dfrac{5.3\ \text{ps}}{\delta\tilde{\nu}/\text{cm}^{-1}}$

(a) $\tau = \dfrac{5.3\ \text{ps}}{0.1}$ = **53 ps**

(b) $\tau = \dfrac{5.3\ \text{ps}}{1}$ = **5 ps**

(c) $\lambda_{\text{obs}} = \dfrac{2.998\times10^{8}\ \text{m s}^{-1}}{1.0\times10^{9}\ \text{s}^{-1}}$ = 30.0 cm

$\tau = \dfrac{5.3\ \text{ps}}{3.33\times10^{-2}\ \text{cm}^{-1}}$ = **159 ps**

11.19 $\delta\tilde{\nu} = \dfrac{5.3\ \text{cm}^{-1}}{\tau/\text{ps}}$

(a) $\tau \approx 1\times10^{-13}$ s = 0.1 ps, implying that $\delta\tilde{\nu}$ = **$50\ \text{cm}^{-1}$**

(b) $\tau \approx 200 \times (1 \times 10^{-13}\ \text{s}) = 20\ \text{ps}$, implying that $\delta\tilde{\nu}$ = **0.27 cm^{-1}**

11.20 $\omega = \left(\frac{k}{\mu}\right)^{1/2}$, so $k = \mu\omega^2 = 4\pi^2\mu c^2 \tilde{\nu}^2$

$$\mu(\text{HF}) = \frac{1.0078 \times 18.9908}{1.0078 + 18.9908}\text{u} = 0.9570\ \text{u}$$

$$\mu(\text{H}^{35}\text{Cl}) = \frac{1.0078 \times 34.9688}{1.0078 + 34.9688}\text{u} = 0.9796\ \text{u}$$

$$\mu(\text{H}^{81}\text{Br}) = \frac{1.0078 \times 80.9163}{1.0078 + 80.9163}\text{u} = 0.9954\ \text{u}$$

$$\mu(\text{H}^{127}\text{I}) = \frac{1.0078 \times 126.9045}{1.0078 + 126.9045}\text{u} = 0.9999\ \text{u}$$

Using the above equation draw up the following table:

	HF	HCl	HBr	HI
$\tilde{\nu}$	4141.3	2988.9	2649.7	2309.5
μ/u	0.9570	0.9796	0.9954	0.9999
k/(N m^{-1})	**967.1**	**515.6**	**411.8**	**314.2**

11.21 Form $\tilde{\nu} = \frac{\omega}{2\pi c} = \frac{1}{2\pi c}\left(\frac{k}{\mu}\right)^{1/2}$ with the values of k from exercise 11.7 and the following reduced masses:

$$\mu(^2\text{HF}) = \frac{2.0141 \times 18.9908}{2.0141 + 18.9908}\text{u} = 1.8210\ \text{u}$$

and similarly for the other halides. Draw up the following table:

	^{2}HF	^{2}HCl	^{2}HBr	^{2}HI
$\tilde{\nu}$	**3002**	**2144**	**1886**	**1640**
μ/u	1.8210	1.9044	1.9652	1.9826
k/(N m^{-1})	967.1	515.6	411.8	314.2

11.22 The molecule is centrosymmetric and so the exclusion rule applies. The mode is **infrared inactive** (symmetric breathing leaves the molecular dipole moment unchanged at zero), and therefore the mode may be **Raman active**.

11.23 $\log \frac{I'}{I} = -\varepsilon[J]l$

$$= -743\ M^{-1}\ cm^{-1} \times (3.25 \times 10^{-3}\ M) \times 0.25\ cm$$

Hence I'/I = 0.249 and the reduction in intensity is 75 percent.

11.24 $\varepsilon = -\frac{1}{[J]l}\log\frac{I'}{I}$

$$= -\ \frac{\log 0.715}{4.33 \times 10^{-4}\ mol\ L^{-1} \times 0.25\ cm}$$

$$= 1350\ dm^3\ mol^{-1}\ cm^{-1} = 1350 \times 10^3\ cm^3\ mol^{-1}\ cm^{-1}$$

$$= \mathbf{1.35 \times 10^6\ cm^2\ mol^{-1}}$$

11.25 $[J] = -\frac{1}{\varepsilon l}\log\frac{I'}{I}$

$$= \frac{-\log 0.602}{291\ L\ mol^{-1}\ cm^{-1} \times 0.65\ cm} = \mathbf{1.16 \times 10^{-3}\ mol\ L^{-1}}$$

11.26 The weak absorption at 30 000 cm^{-1} is typical of a carbonyl group. The strong C=C absorption, which typically occurs at about 180 nm, has been shifted to longer wavelength (213 nm) because of the double bond and the CO group.

11.27 $\varepsilon = -\frac{1}{[\mathrm{J}]l}\log\frac{I'}{I}$ with l = 0.20 cm

Use this formula to draw up the following table:

$[Br_2]$/mol L^{-1}	0.0010	0.0050	0.0100	0.0500
I'/I	0.814	0.356	0.127	3.0×10^{-5}
$\varepsilon/(M^{-1}\ cm^{-1})$	447	449	448	452

The average absorption coefficient is ε = **450 mol^{-1} L cm^{-1}**

11.28 $\varepsilon = -\frac{1}{[\mathrm{J}]l}\log\frac{I'}{I}$

$$= \frac{-1}{0.010\ \mathrm{M} \times 0.20\ \mathrm{cm}}\log 0.48 = \mathbf{160\ mol^{-1}\ L\ cm^{-1}}$$

$$\frac{I'}{I} = 10^{-[\mathrm{J}]\varepsilon l} = 10^{-0.010\ \mathrm{M}\ \times(159\ \mathrm{M}^{-1}\ \mathrm{cm}^{-1})\times 0.40} = 0.23, \text{ or } \mathbf{23\ percent}$$

11.29 $l = -\frac{1}{[\mathrm{J}]\varepsilon}\log\frac{I'}{I}$

For water, $[H_2O] \approx \frac{1.00\ \mathrm{kg}}{18.02\ \mathrm{g\ mol^{-1}}} = 55.5$ M

$\varepsilon[\mathrm{J}] = 55.5\ \mathrm{M} \times (6.2 \times 10^{-5}\ \mathrm{M^{-1}\ cm^{-1}}) = 3.4 \times 10^{-3}\ \mathrm{cm^{-1}}$

$= 0.34\ \mathrm{m^{-1}}$, so $1/(0.34) = 2.9$ m

$$l = -2.9\text{ m} \times \log\frac{I'}{I}$$

(a) $I'/I = 0.5$, $l = -2.9\text{ m} \times \log 0.5 =$ **0.9 m**
(b) $I'/I = 0.1$, $l = -2.9 \times \log 0.1 =$ **3 m**

11.30 The initial energy of the photons minus the kinetic energy will give the ionization energy. Therefore, the ionization energies are: **10.21 eV, 12.99 eV, and 16.00 eV.** The lowest ionization energy, highest kinetic energy probably results from a nonbonding orbital, the other two ionizations are probably bonding orbitals so a configuration like $2p\pi^2 2p\pi^{*1}$ would give the spectrum.

11.31 $A = c_1\varepsilon_1 + c_2\varepsilon_2$

$0.660 = c_1(2.00 \times 10^3) + c_2(1.12 \times 10^4)$

$0.221 = c_1(5.40 \times 10^3) + c_2(1.5 \times 10^5)$

Using these two equation and solving for $c_1 =$ **2.11×10^{-5} M** and $c_2 =$ **7.14×10^{-5} M**

11.32 For identical nuclei with spin 1/2, there will be N+1 lines from the splitting. In this case 8 lines. The lines will have intensities of
1:7:21:35:35:21:7:1. The intensities can be determined from the Pascal triangle of expanding $(1+x)^N$ and taking the coefficients.

11.33 Since each resonance is split into three lines by a single N nuclei, the result will be:
(a) **quintet 1:2:3:2:1** and (b) **heptet 1:3:6:7:6:3:1**

11.34 $E = -g_I\mu_N m_I B$ with $m_I = \frac{3}{2}, \frac{1}{2}, -\frac{1}{2}, -\frac{3}{2}$ $[1, \gamma\hbar = g_I\mu_N]$

$= -0.4289 \times (5.051 \times 10^{-27}\ \mathrm{J\ T^{-1}}) \times (7.500\ \mathrm{T}) \times m_I$

$=$ **$\underline{-1.625 \times 10^{-26}\ \mathrm{J} \times m_I}$**

11.35 $|DE| = |-g_I\mu_N B(-1\ -1)|$ $[I = 1, m_I(\max) = +1, m_I(\min) = -1]$

$= 2\,g_I\mu_N B$

$= 2 \times 0.4036 \times (5.051 \times 10^{-27}\ \mathrm{J\ T^{-1}}) \times 15.00\ \mathrm{T}$

$= 6.116 \times 10^{-26}\ \mathrm{J}$

$2\,g_I\mu_N B = hn = 6.116 \times 10^{-26}\ \mathrm{J}$

$n = (6.116 \times 10^{-26}\ \mathrm{J})/(6.626 \times 10^{-34}\ \mathrm{J\ Hz}) =$ **$\underline{92.3\ \mathrm{MHz}}$**

11.36 $B = \dfrac{h\nu}{g_I\mu_N}$

$$= \frac{(6.626 \times 10^{-34}\ \mathrm{J\,Hz^{-1}}) \times (550.0 \times 10^{6}\ \mathrm{Hz})}{5.586 \times (5.051 \times 10^{-27}\ \mathrm{J\,T^{-1}})} = \mathbf{\underline{12.92\ T}}$$

11.37 $B_{loc} = (1 - s)B$

$|DB_{loc}| = |(Ds)|B \approx |\{d(CH_3) - d(CHO)\}|B$

$= |(2.20 - 9.80)| \times 10^{-6}\ B = (7.60 \times 10^{-6})B$

(a) $B = 1.5\ \mathrm{T}$, $|DB_{loc}| = (7.60 \times 10^{-6}) \times 1.5\ \mathrm{T} =$ **$\underline{11\ \mathrm{mT}}$**

(b) $B = 6.0\ \mathrm{T}$, $|DB_{loc}| = (7.60 \times 10^{-6}) \times 6.0\ \mathrm{T} =$ **$\underline{46\ \mathrm{mT}}$**

11.38 $|D\nu| = |\Delta\delta| \times \nu_o = 7.60 \times 10^{-6} \times \nu_o$

(a) ν_o = 300 MHz, $|\Delta\nu| = (7.60 \times 10^{-6}) \times 300$ MHz = **2.28 kHz**

(b) ν_o = 550 MHz, $|\Delta\nu| = (7.60 \times 10^{-6}) \times 550$ MHz = **4.18 kHz**

11.39 (a) The spectrum is shown below

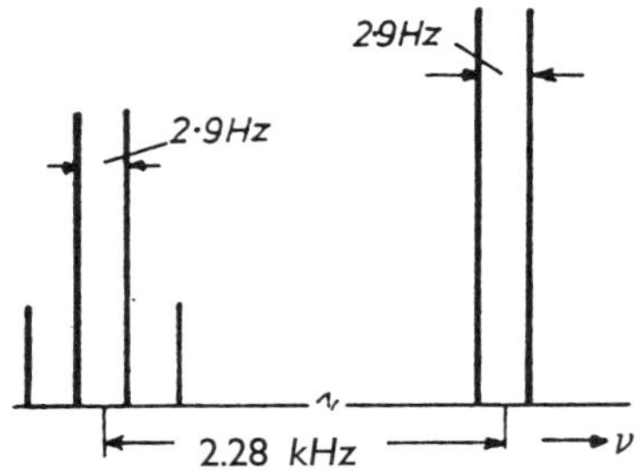

(b) When the frequency is changed to 550 MHz, the separation of the CH_3 and CHO resonance increases (4.18 kHz) the fine structure remains unchanged, and the intensity increases.

11.40 The four equivalent ^{19}F nuclei (I = 1/2) give a single line. However, the ^{10}B nucleus (I = 3, 19.6 percent abundant) results in 2 x 3 + 1 = 7 lines and the ^{11}B nucleus (I = 3/2, 80.4 percent abundant) results in 2 x 3/2 + 1 = 4 lines. The splitting arising from the ^{11}B nucleus will be larger, by a factor of 1.5. Moreover the total intensity of the 4 lines will be greater due to the greater abundance. The individual line intensities will be in the ratio of 8:1 (half the number of lines, and four times as abundant). The spectrum is shown below:

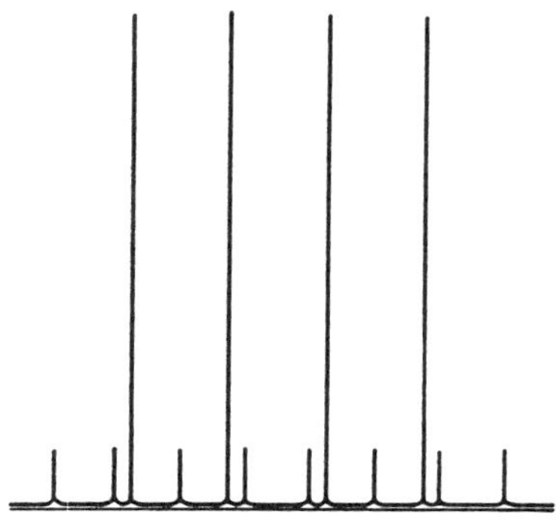

11.41 The A, M, and X resonances lie in distinctively different groups. The A resonance is split into a 1:2:1 triplet by the M nuclei, and each line of that triplet is split into a 1:4:6:4:1 quintet by the X nuclei. The M resonance is split into a 1:3:3:1 quartet by the A nuclei and each line is split into a quintet by the X nuclei. The X resonance is split into a quarte by the A nuclei and then each line is split into a triplet by the M nuclei. The spectrum is shown below:

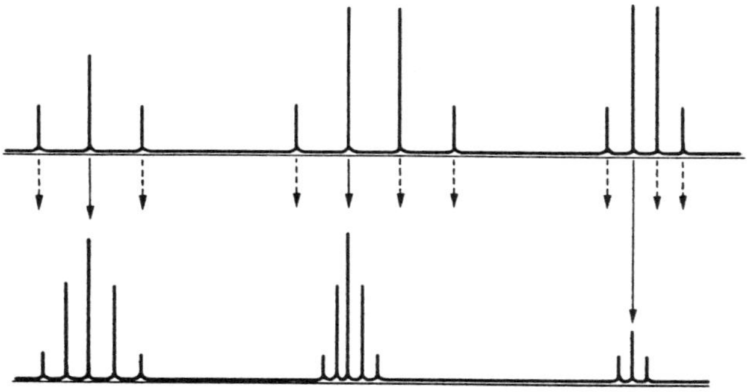

Additional problems

11.1 A biological sample in a 1 cm pathlength cell absorbs 80% of the incident light. The molar absorption coefficient is 4.2 x 10^4 $mol^{-1}L\ cm^{-1}$. What is the concentration of the sample?

11.2 An unknown sample has an absorbance of 1.8 in a 1 cm pathlength cell. What fraction of light is transmitted?

11.3 A railroad sign has green lights (520 nm). What wavelength will the lights appear to an engineer approaching at 1000 m s^{-1}?

11.4 A titanium filament emits light at 700 nm when heated. The spectral line width at half-height is 3 pm. What is the temperature of the filament?

11.5 Which of the following molecules will have rotational absorption spectra: (a) CH_3CHCl_2 (b) H_2S (c) CH_3CH_3 (d) CH_2O ?

11.6 Why is spectral line broadening less when working at low pressures?

11.7 What is the nmr shift of the resonance from TMS of a nuclei with d = 2.50 in a spectrometer operating at a frequency of 500 MHz?

11.8 Describe the nmr peak splitting and intensities for the H in CH_3CH_2OH.

11.9 What fine structure can be expected for the protons in $^{13}CH_4$?

11.10 Calculate the ionization energy of a molecule whose photoelectrons are ejected with a velocity of 1.8 x 10^3 km s^{-1} by radiation from a helium discharge lamp (21 eV).

11.11 Describe the proton nmr spectrum of $CH_3CClH\text{-}CO_2H$.

Answers to additional problems

Chapter 1

1.1 T = 200 K **1.2** (a) 24.5 atm (b) 24.1 atm **1.3** 0.082 mol **1.4** 2.5 atm **1.5** 1.23 g L^{-1} **1.6** 5.4×10^{-2} m **1.7** (a) 5.0×10^4 (b) 3.5×10^4 (c) 2.7×10^5 **1.8** (a) 4.1×10^5 L (b) 4.1×10^5 L **1.9** 0.207 atm **1.10** 241 g mol^{-1} **1.11** (a) 66.8 nm (b) 6.68 nm **1.12** N_2 (a) 36 kPa (b) 145 kPa; O_2 (a) 48 kPa (b) 190 kPa; Ar (a) 16 kPa (b) 64 kPa **1.13** $N_{2,\,p}$ = 22 kPa, CO_2, p = 44 kPa, O_2, p = 24 kPa **1.14** 0.017

Chapter 2

2.1 -4.0 kJ **2.2** (a) -20 kJ (b) 0 kJ **2..3** +2.5 kJ **2.4** (a) 180 J (b) 180 J **2.5** 8.4×10^4 kJ **2.6** 19 J K^{-1} g^{-1} **2.7** -75.2 kJ mol^{-1} **2.8** (a) endothermic (b) exothermic (c) exothermic (d) exothermic **2.9** -221 kJ **2.10** 2.1×10^4 **2.11** (a) 203 kJ (b) 0 kJ **2.12** -1453 kJ = -726 kJ mol^{-1} **2.13** -850 kJ **2.14** q = ΔH = 1.8 kJ, ΔU = 1.3 kJ

Chapter 3

3.1 358 J K^{-1} **3.2** -424 J K^{-1} **3.3** ΔH (+), ΔS (+) **3.4** ΔH = -2.29 kJ, ΔS = -98.74 J K^{-1}, **3.5** 3.40 ln 1/10 J K^{-1} **3.6** 180 J K^{-1} **3.7** K = 10^{261} **3.8** 448 K **3.9** ΔG^{θ} = -56 kJ **3.10** ΔS = 0.96 J K^{-1} **3.11** +43.2 J K^{-1} **3.12** $n(R-C_V)\ln 2$

Chapter 4

4.1. 93.4 Torr **4.2** 23.4 Torr **4.3** 100.4° C **4.4** 110 g mol^{-1} **4.5** 157 g **4.6** The critical temperature of ammonia is above room temperature whereas the critical temperature of nitrogen is below room temperature. **4.7** 80° C **4.8** (a) gas goes to liquid at high pressure (b) gas goes to solid at pressure above 5.1 atm (c) the sample will remain as gas at all pressures. **4.9** 0.30 M **4.10** Increasing the pressure raises the boiling temperature and at high altitude the boiling temperature is low due to low atmospheric pressure. **4.11** As the pressure decreases initially the water boils. Evaporation is endothermic so the water will freeze leaving ice. At further low pressure the

ice sublimes. **4.12** 0.16 mol L^{-1}, 1.2×10^{-5} mol L^{-1}
4.13 Osmotic pressure. Change in temperature too small to measure. **4.14** (a) 5.1×10^{-4} mol kg^{-1} (b) 6.51×10^{-4} mol kg^{-1}

Chapter 5

5.1 2.2×10^{-5} **5.2** p/p^{θ}, $p = p(H_2O)$ **5.3** 5.00×10^{-13}
5.4 2.0×10^{-5} **5.5** 0.013 **5.6** 0.0018 **5.7** pH = 10.7
5.8 acidic, basic, acidic **5.9** $K_s = 6.9 \times 10^{-9}$
5.10 increase

Chapter 6

6.1 $2(MnO_4^- + 5e^- + 8H^+ \rightarrow Mn^{2+} + 4H_2O)$ $E^{\theta} = 1.51$ V
$5(ClO_3^- + H_2O \rightarrow ClO_4^- + 2H^+ + 2e^-)$ $-E^{\theta} = -1.23$V
E^{θ} cell = 0.28 V **6.2** 2.9 kJ mol^{-1} **6.3** a) $E < E^{\theta}$ b) $E > E^{\theta}$
6.4 $\Delta G^{\circ} = -362$ kJ, $K = 4.9 \times 10^{63}$ **6.5** $E^{\theta} = +2.12$ V **6.6** pH increases $E > E^{\theta}$ **6.7** E = 1.2 V **6.8** pH = 1.0 **6.9** 1.75×10^{-6}
6.10 pH increase will increase the reaction **6.11** 4.7×10^{25}
6.12 $E^{\theta} = +2.12$ V, $\Delta G^{\theta} < 0$, therefore spontaneous.

Chapter 7

7.1 1.7×10^5 J mol^{-1} **7.2** $k_2 = 3.2 \times 10^{-2}$ L mol^{-1} s^{-1}
7.3 1.5×10^5 s **7.4** $k[H_2][I_2]$ = rate **7.5** zero order with regard to A **7.6** first order **7.7** $k = M^{1/2}$ s^{-1} **7.8** 5.3×10^4 J mol^{-1}
7.9 75 s, $k = 9.2 \times 10^{-3}$ **7.10** (a) mol L^{-1} s^{-1} (b) rate = k, mol L^{-1} s^{-1} (c) L^2 mol^{-2} s^{-1} **7.11** yes, 1st step is rate limiting
7.12 noncompetitive

Chapter 8

8.1 2.96×10^{-19} J **8.2** 486 nm **8.3** 5.41×10^{14} s^{-1}
8.4 3.10 pm **8.5** 5500 K **8.6** (a) 1.10×10^{-27} kg m s^{-1}
(b) 6.6×10^{-23} kg m s^{-1} **8.7** 7.27×10^{-11} m **8.8** 6.4×10^4 m s^{-1}
8.9 (a) 2.15×10^{-6} pm^{-3} (b) 3.9×10^{-8} pm^{-3} **8.10** $\psi^2\,\delta\tau$ gives the probability of finding the electron in the region $\delta\tau$ at that point.

Chapter 9

9.1 P-H, C-Cl, H-O **9.2** N_2, Bond order = 3 **9.3** T-shaped, 2 lone pairs **9.4** H_2O (2 lone pairs) **9.5** Ne_2, B.O. = 0, unstable; P_2 B.O. = 3, stable **9.6** O_2^+ : $\sigma_{2s}^2\sigma_{2s}^{*2}\pi_{2p}^4\sigma_{2p}^2\pi_{2p}^{*1}$ B.O. = 2.5; O_2^- : $\sigma_{2s}^2\sigma_{2s}^{*2}\pi_{2p}^4\sigma_{2p}^2\pi_{2p}^{*3}$ B.O. = 1.5 **9.7** a) OF_2 angular, sp^3 b) CF_4 Tetrahedral, sp^3 c) KrF_2 Linear dsp^3 **9.8** π_{2p}^* orbitals will fill singly first. It takes energy to pair up the electrons.
9.9 Double bonded C are sp^2, third C is sp^3 **9.10** No.

Chapter 10

10.1 233 pm **10.2** 194.5 g mol^{-1} **10.3** body centered
10.4 128 pm **10.5** 786 kJ mol^{-1} **10.6** linear
10.7 2.5×10^{-7} m^2 s^{-1} **10.8** pH 5.75 **10.9** 6.2×10^2 hr
10.10 trans Cl-CH=CH-Cl

Chapter 11

11.1 1.7×10^{-5} mol L^{-1} **11.2** 0.016 **11.3** 0.999997 x 520 nm
11.4 1700 K **11.5** (a),(b),(d) **11.6** At low pressure collisional deactivation is minimized. **11.7** 1.25 kHz **11.8** H from OH singlet; CH_3 triplet 1:2:1; CH_2 quartet 1:3:3:1. **11.9** 1:1:1 triplet from ^{13}C splitting **11.10** 12 eV **11.11** CH_3 doublet 1:1 ; CClH quartet 1:3:3:1; CO_2H singlet.